The Nature of the Firm in the Oil Industry

“There is an abundance of theorising about global value chains, but painfully little in-depth analysis of real world examples. Dr. Beyazay has undertaken a unique, path breaking study of the value chain of the oil and gas industry. It provides a remarkable insight into the relationship between the oil majors and their key suppliers in the upper reaches of the value chain. It deserves to be widely read by scholars, business practitioners and policy-makers.”

—Professor Peter Nolan, Director of the Centre of Development Studies, University of Cambridge

“The functioning of the global oil industry hinges on the relationships between different types of firms, and especially between oil services companies and the private or state-owned oil companies that hire them. This book offers a rigorous and current analysis of these relationships and how they are evolving over time. I highly recommend it to anyone interested in how today’s oil industry actually works.”

—Mark Thurber, coeditor and contributor to Oil and Governance; *research scholar at Stanford University*

Firm-to-firm relationships, along with the overall structure of industries, have changed markedly over the past decades. Replacing the model of vertical integration with one of global business, firms have started to outsource more by using a wider global network. At the same time, they have begun to increase their control and coordination along the value chain to remain competitive, blurring the boundaries between companies. Understanding the changing nature of the firm and its role in coordinating the supply chain will help firms to better define global competitive strategies.

The Nature of the Firm in the Oil Industry aims to facilitate the understanding of ‘the firm’ in the oil industry via the analysis of the specific relationship between international oil companies, which are among the world’s biggest firms and which act as ‘core system integrators’, and the oil services companies, which help to find, extract, produce and distribute oil along the petroleum industry supply chain. This relationship serves as an example of deep integration by core system integrators and provides insights into the change in the nature of the firm in the era of modern globalization. Aimed at researchers and academics, *The Nature of the Firm in the Oil Industry* offers a thorough examination of this relationship to shed light on the nature of the firm, both in the oil industry and in global business today. It is a humble attempt to better understand the firm in a crucial industry.

Basak Beyazay-Odemis received her PhD and her MPhil degree at the University of Cambridge in the Centre of Development Studies. Prior to her studies in Cambridge, she obtained a double diploma of ENA (Promotion Copernic) and a Master in Public Administration at the Ecole Nationale d’Administration (ENA) in France, and a BA in the Francophone Department of Political & Administrative Sciences of Marmara University in Turkey. Her principal areas of research include energy markets, international oil companies, the oil industry and the economies and policies of developing countries, including her native country, Turkey. She currently works as an energy trader at an international oil and gas company in London.

Routledge Studies in International Business and the World Economy

For a full list of titles in this series, please visit www.routledge.com

34 Governing Interests
Business associations facing internationalization
Edited by Wolfgang Streeck, Jürgen Grote, Volker Schneider and Jelle Visser

35 Infrastructure Development in the Pacific Region
Edited by Akira Kohsaka

36 Big Business and Economic Development
Conglomerates and economic groups in developing countries and transition economies under globalisation
Edited by Alex E. Fernández Jilberto and Barbara Hogenboom

37 International Business Geography
Case studies of corporate firms
Edited by Piet Pellenbarg and Egbert Wever

38 The World Bank and Global Managerialism
Jonathan Murphy

39 Contemporary Corporate Strategy
Global perspectives
Edited by John Saee

40 Trade, Globalization and Poverty
Edited by Elias Dinopoulos, Pravin Krishna, Arvind Panagariya and Kar-yiu Wong

41 International Management and Language
Susanne Tietze

42 Space, Oil and Capital
Mazen Labban

43 Petroleum Taxation
Sharing the oil wealth: a study of petroleum taxation yesterday, today and tomorrow
Carole Nakhle

44 International Trade Theory
A critical review
Murray C. Kemp

45 The Political Economy of Oil and Gas in Africa
The case of Nigeria
Soala Ariweriokuma

46 Successfully Doing Business/Marketing in Eastern Europe
Edited by V.H. Manek Kirpalani, Lechoslaw Gabarski and Erdener Kaynak

47 Marketing in Developing Countries: Nigerian Advertising and Mass Media
Emmanuel C. Alozie

48 Business and Management Environment in Saudi Arabia: Challenges and Opportunities for multinational corporations
Abbas J. Ali

49 International Growth of Small and Medium Enterprises
Edited by Niina Nummela

50 Multinational Enterprises and Innovation
Regional Learning in Networks
Martin Heidenreich, Christoph Barmeyer, Knut Koschatzky, Jannika Mattes, Elisabeth Baier, and Katharina Krüth

51 Globalisation and Advertising in Emerging Economies
Brazil, Russia, India and China
Lynne Ciochetto

52 Corporate Social Responsibility and Global Labor Standards
Firms and activists in the making of private regulation
Luc Fransen

53 Corporations, Global Governance and Post-Conflict Reconstruction
Peter Davis

54 Self-Initiated Expatriation
Individual, organizational, and national perspectives
Edited by Maike Andresen, Akram Al Ariss and Matthias Walther

55 Marketing Management in Asia
Edited by Stan Paliwoda, Tim Andrews and Junsong Chen

56 Retailing in Emerging Markets
A policy and strategy perspective
Edited by Malobi Mukherjee, Richard Cuthbertson and Elizabeth Howard

57 Global Strategies in Retailing
Asian and European experiences
Edited by John Dawson and Masao Mukoyama

58 Emerging Business Ventures under Market Socialism
Entrepreneurship in China
Ping Zheng and Richard Scase

59 The Globalization of Executive Search
Professional Services
Strategy and Dynamics in the Contemporary World
Jonathan V. Beaverstock, James R. Faulconbridge and Sarah J.E. Hall

60 Born Globals, Networks and the Large Multinational Enterprise
Insights from Bangalore and beyond
Shameen Prashantham

61 Big Business and Brazil's Economic Reforms
Luiz F. Kormann

62 The Nature of the Firm in the Oil Industry
International Oil Companies in Global Business
Basak Beyazay-Odemis

The Nature of the Firm in the Oil Industry

International Oil Companies in Global Business

Basak Beyazay-Odemis

NEW YORK AND LONDON

First published 2016
by Routledge
711 Third Avenue, New York, NY 10017

and by Routledge
2 Park Square, Milton Park, Abingdon, Oxon OX14 4RN

First issued in paperback 2018

Routledge is an imprint of the Taylor & Francis Group, an informa business

Library of Congress Cataloging-in-Publication Data
Beyazay-Odemis, Basak, author.
The nature of the firm in the oil industry : international oil companies in global business / by Basak Beyazay-Odemis.
pages cm. — (Routledge studies in international business and the world economy ; 62)
Includes bibliographical references and index.
1. Petroleum industry and trade. 2. Industrial organization. I. Title.
HD9560.5.B467 2015
338.8′87223382—dc23 2015027902

ISBN 13: 978-1-138-34067-1 (pbk)
ISBN 13: 978-1-138-82684-7 (hbk)

Typeset in Sabon
by Apex CoVantage, LLC

Disclaimer

The book is entirely independent from any company. The views expressed in the book are written solely in the author's private capacity and do not in any way represent the views of her employer or any other company. Responsibility for the opinions expressed in the book rests entirely with the author.

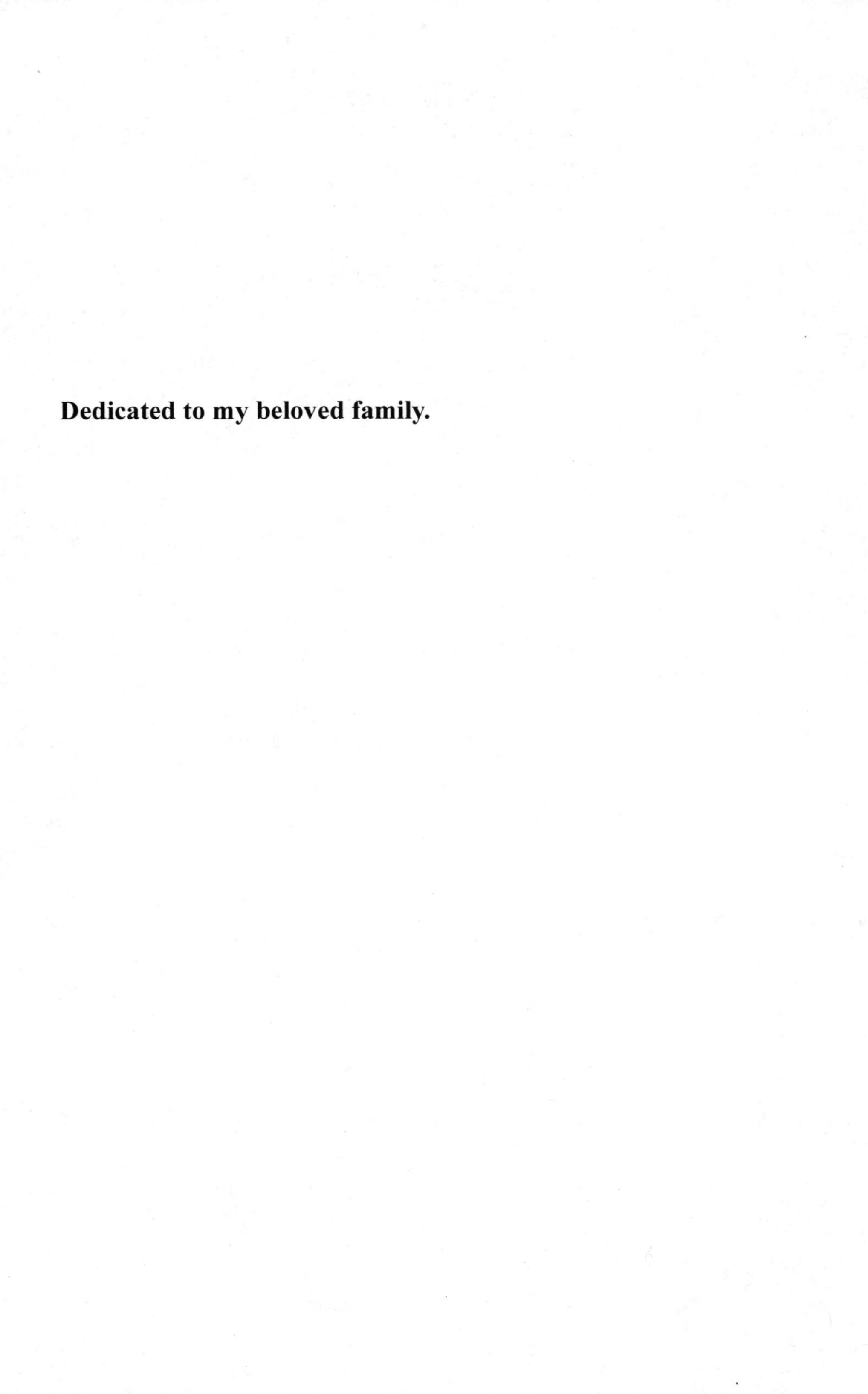

Dedicated to my beloved family.

Contents

Tables and Figures

TABLES

FIGURES

Abbreviations

$/USD: United States Dollar
€/EUR: European Union Euro
3D: Three Dimensional
4D: Four Dimensional
BBL: Oil Barrel
BHI: Baker Hughes International
BN: Billion
BOP: Blowout Preventer
CAM: Cameron International Corporation
CAPEX: Capital Expenditure
CEO: Chief Executive Officer
CFO: Chief Financial Officer
CONOCO: ConocoPhillips
COSL: China Oilfield Services Limited
EBITDA: Earnings Before Interest, Taxes, Depreciation and Amortization
EFA: Enterprise Frame Agreement
EPC: Engineering, Procurement and Construction
EXXON: ExxonMobil Corporation
E&P: Exploration and Production
FID: Final Investment Decision
FPS: Floating Production System
FPSO: Floating Production, Storage and Offloading
FT: Financial Times
HAL: Halliburton
HPHT: High Pressure, High Temperature
HSE: Health, Safety and Environment
IFP: Institut Français Du Pétrole (French Institute of Petroleum)

INOC:	Iraq National Oil Company
IOC:	International Oil Company
IPB:	Integrated Production Bundle
IT:	Information Technology
JV:	Joint Venture
L12M:	Last 12 Months
LBM/GAL:	Pound Mass per Gallon
LNG:	Liquefied Natural Gas
LPG:	Liquefied Petroleum Gas
LWD:	Logging While Drilling
M&A:	Mergers & Acquisitions
MAZ:	Multi Azimuth
MN:	Million
MPA:	Megapascal Pressure Unit
NIOC:	National Iranian Oil Company
NNPC:	Nigerian National Petroleum Corporation
NOC:	National Oil Company
OM:	Oil Major
OSC:	Oil Services Company
PDVSA:	Petróleos de Venezuela
PEMEX:	Petróleos Mexicanos
PIW:	Petroleum Intelligence Weekly
PSA:	Production Share Agreement
PSI/FT:	1 Pound per Square Inch/Foot
RIG:	Transocean
R&D:	Research & Development
ROE:	Return on Equity
RFP:	Request for Proposal
SHELL:	Royal Dutch Shell
SLB:	Schlumberger
SURF:	Subsea Umbilicals, Risers and Flowlines
TFLOP:	Teraflop
TR:	Técnicas Reunidas
US:	The United States
UK:	The United Kingdom
WAZ:	Wide Azimuth
WI:	Weatherford International
WSJ:	Wall Street Journal

Acknowledgments

This book is based on my doctoral dissertation. Many people have helped me during the writing of this book and the doctoral work which formed the basis for it. I am first and foremost indebted to my research supervisor, Professor Peter Nolan, for our innumerable discussions, his support and his careful reading of the thesis. He gave me room to develop my research in freedom, while providing me with valuable guidance. I am also grateful to the Centre of Development Studies at Cambridge University for allowing me to benefit from its valuable academic team during my masters and PhD degree work.

This study would not be what it is now without the people I interviewed while preparing my thesis. I thank them for the time and trust they accorded me. I benefited tremendously from the richness of information they provided and their insightful observations. I am also grateful to Total, which I had the pleasure to work for, for allowing me to conduct my PhD and write this book. I should, however, specify that both the thesis and the book are not sponsored by Total and are entirely independent from any company. Responsibility for the opinions expressed in the work rests entirely with me.

I am grateful to my mum and dad, who taught me the virtues of hard work and perseverance, and to my little sister, Burcu Beyazay, for her constant support in every aspect of my life. I thank my close friends, family and extended Odemis and Kirdok families for respecting my choices and giving me unconditional love and support.

I am grateful to my daughters, Melis and Deniz Odemis, for allowing me to continue working after their birth by being so well behaved and sleeping through the night. Finally, my biggest thanks go to my husband, the love of my life, Ozgur Odemis, for his emotional and intellectual support. Without him, the thesis and the book could not have been written. He is crucial to shaping my ideas and making my life more joyful and meaningful.

Introduction

The nature of the firm has changed in the globalized world. This change is mainly due to the transformation in the overall structure of industries and the corresponding evolution in firm-to-firm relationships over the past decades.

During the 'global business revolution' of the 1990s, giant firms were created as a result of industrial consolidation driven by multiple mergers and acquisitions. At the same time, the value chain in each sector evolved and expanded due to the disintegration of previously vertically integrated companies. Large global firms focusing on their 'core business' replaced the previous model of vertically integrated firms. In this environment, the decision on which activities to manage and produce in-house and which activities to outsource became central to a firm's ability to remain competitive. In a rapidly globalizing world, more and more decisions were made in favour of outsourcing as advances in communication and technology greatly simplified the coordination of activities across national boundaries. Firms therefore have started to outsource more by using a wider global network. At the same time, they have begun to increase their control and coordination along the value chain to remain competitive. These trends have changed the nature of the firm and blurred its boundaries. The large corporates started acting as the 'system integrators' that plan, coordinate and integrate knowledge across the whole supply chain (Nolan 2000). These firms began coordinating a vast area of business activity outside the boundaries of legal entity in terms of ownership (Nolan et al. 2005). Understanding the changing 'nature' of the firm and its role in coordinating the supply chain will help firms to better define global competitive strategies.

This book is a humble attempt to understand the nature of the firm in the oil[1] industry. It has a particular focus on international oil companies (IOCs), which are among the world's biggest firms that work on a vital nonmanufactured product—oil, which is used in everything from electricity generation to automobiles, to asphalt and road oil, to plastics and cosmetics. Because this product is found in nature, the oil industry supply chain involves the oil services companies (OSCs) helping to find, extract, produce and distribute oil rather than companies producing components of a manufactured product, as is typical of industrial sectors such as the automobile sector. Still, similar to other industries, the oil industry has moved from a vertically integrated model to an outsourcing-based model in which IOCs contract OSCs and their subcontractors to explore and produce oil. OSCs are the firms that supply the technology required to acquire reservoir-related data, to drill wells and to produce oil. Because oil is increasingly being sourced from challenging areas which demand the expertise of these technology suppliers, the relationship between IOCs and OSCs is changing. It is therefore timely to analyse the transformation in the oil industry that is occurring as a result of vertical disintegration and an evolving structure.

The aim of this book is to capture the transformation of the oil industry supply chain and the impact of this transformation on the nature of the firm. The main objective is to find out how and why the nature of the firm and the structure of the value chain in the petroleum industry have changed, through the analysis of the relationship between IOCs and OSCs. The goal is to discover the determinants behind the transformation and the resulting implications for the firm and the value chain.

The oil industry is of course a huge topic that can fruitfully give rise to a multitude of books and studies. This book does not aspire to examine every aspect of the oil industry supply chain but aims to shed light on the transformation in the oil industry value chain due to the global business revolution by developing a deeper understanding of the prospective coordination and planning role of the core firm in the oil industry.

The book will proceed as follows. Chapter 1 will introduce key concepts regarding globalization, the nature of the firm, vertical disintegration and coordination by system integrators. Chapter 2 will portray the oil industry supply chain by going through its upstream and downstream segments and the companies involved. It will also discuss the petroleum supply chain in detail, identifying elements determining its particular characteristics, its differentiating elements and its transformation over time. Chapter 3 will focus on the implications of the changes for IOCs. It will explore the relationship between IOCs and OSCs by examining aspects ranging from procurement to research and development. It will analyse how and why the relationship has changed over time. Chapter 4 completes the previous chapter by illustrating the relationship via three case studies. Finally the concluding Chapter 5 will address the transformation of the nature of the modern firm in the oil industry.

NOTE

1. This book uses the terms *oil companies* and *oil* to facilitate reading. It should, however, be noted that the majority of cited companies are active in all hydrocarbons, including petroleum and natural gas. The information could be considered valid for oil and other hydrocarbons.

1 Global Value Chain & Changes in the Nature of the Firm

1.1 THE NATURE OF THE FIRM

Since the publication of Coase's 1937 essay 'The Nature of the Firm', competing and complementary theories have emerged to explain why individuals might choose to form partnerships or companies rather than trade bilaterally through markets.

Prior to Coase and within the context of price theory, firms were viewed as identical production functions where homogenous inputs were transformed into homogenous outputs. From this perspective, the firm was no more than a theoretical link, explaining the relationship between changes in price and quantity in response to variations in exogenous factors (Langlois and Koppl 1991; Loasby 1976). In reality, the price theory was never intended to be a theory of the firm as an organization or as an institution (Machlup 1967). For this reason, price theory has proved unsatisfactory in its ability to explain the existence, boundaries and internal structure of the firm (Langlois and Foss 1997).

Coase confronted and challenged price theory by highlighting the existence of transaction costs in order to explain the *raison d'être* of the firm. He argued that the main reason for establishing a firm is to avoid the costs of using the price mechanism; in other words, to avoid what he termed 'transaction costs'. According to Coase, it is transaction costs, such as the cost of researching prices, marketing, organizing production and negotiating contracts, that explain how firms are structured and developed. Accordingly, a firm aims to minimize transaction costs and the uncertainty that comes with the use of short-term contracts in an exchange market (Coase 1937).

More recently, the Coasean literature has focused on the choice between firms and markets and has viewed the 'nature' of the firm as fundamentally contractual. That is, firms are bundles of contracts creating and realigning incentives (Langlois and Foss 1997; Williamson 2002). However, complementary theories have emerged, including the resource-based view proposed by Edith Penrose and the capabilities perspective developed by a consortium of researchers including Langlois, Foss and Richardson.

In her 'resource' theory, Penrose (1959) proposed that firms are a collection of productive resources and the disposal of these resources is determined by their differing uses and by administrative decisions. Firms create economic value not only due to the possession of resources but also due to effective and innovative management of these resources. Therefore, optimum growth for a firm involves finding the most profitable balance between the exploitation of existing resources and the development of new ones (Penrose 1959).

A similar theory is proposed by the 'capabilities' perspective, which sees the firm as a repository of productive knowledge with certain nonstandard characteristics, namely 'capabilities'. Capabilities are seen as team-embodied and refer to the partly tacit production and organization of knowledge that can be operated by members of the firm to create economic value. Knowledge, skills and traditions form the distinctive capabilities of a firm and thus determine the firm's boundaries (Langlois and Foss 1997).

The main implication of the 'capabilities view' is that complementarities and similarities of capabilities affect the organizational structure of firms and the firm–market boundary in an economy (Richardson 1998). Capabilities form an independent causal factor behind the patterns of economic organization. The capabilities perspective places the firm at the centre of analysis, instead of the transactions and associated transaction costs. This emphasizes the specific nature of the firm's facilities and skills as the most significant factor in determining what will be done by the firm and by the market (Chandler 1992).

As a surprising addition to his previous work, Coase also argued in the early 1990s that transaction costs are not the only factor influencing the formation of a firm (Coase 1990).

> While transaction cost consideration undoubtedly explains why firms come into existence, once most production is carried out within firms and most transactions are firm-firm transactions, not factor-factor transactions, the level of transaction cost will be greatly reduced and the dominant factor in determining the institutional structure of production will in general no longer be transaction costs but the relative costs of different firms in organizing particular activities.
>
> (Coase, as quoted in Langlois and Foss 1997, 19)

The function whereby firms plan and organize the activities of other economic actors is referred to as the 'coordination' function. While many competing and complementary theories seek to explain the existence of firms, the main function of the firm that will be used here to facilitate discussion of the nature of firms in the oil industry is precisely this coordination function.

A close reading of Coase's article on the nature of the firm suggests that his explanation for the emergence of the firm is ultimately a planning and coordination one (Langlois and Foss 1997).

> Owing to the risk attitude of the people concerned, they may prefer to make a long rather than a short-term contract. Now, owing to the

> difficulty of forecasting, the longer the period of the contract is for the supply of the commodity or service, the less possible, and indeed, the less desirable it is for the person purchasing to specify what the other contracting party is expected to do. It may well be a matter of indifference to the person supplying the service or commodity which of several courses of action is taken, but not to the purchaser of that service or commodity. But the purchaser will not know which of these several courses he will want the supplier to take. Therefore, the service which is being provided is expressed in general terms, the exact details being left until a later date. All that is stated in the contract is the limits to what the persons supplying the commodity or service is expected to do. The details of what the supplier is expected to do is not stated in the contract but is decided later by the purchaser, When the direction of resources (within the limits of the contract) becomes dependent on the buyer in this way, that relationship which I term a 'firm' may be obtained. A firm is likely therefore to emerge in those cases where a very short term contract would be unsatisfactory.
>
> (Coase 1937, 391–92)

Coase also argues that firms continue to exist only if they perform their coordination function at a lower cost than could be achieved by market transactions and also at a lower cost than that of other firms. To have an efficient economic system, it is necessary to have not only markets but also areas of planning within appropriately sized organizations (Coase 1992).

1.2 CHANGES IN THE NATURE OF THE FIRM

Vertical Disintegration

As discussed above, the emergence of the firm is linked to necessities of planning and coordination. However, the coordination role of the firm is generally seen as the coordination of production factors within a firm. Nevertheless, as a result of the changes in the structure of the value chain (mainly vertical disintegration), firms have increased their coordination function outside the firm's legal boundaries and have begun coordinating the activities of companies along the entire value chain (Nolan et al. 2002).

Historically, the continuing growth of modern industrial enterprise arose via horizontal combination, vertical integration (primarily to assure supplies) and movement into new markets. The strategy of backward integration reflected the need to ensure a steady flow of materials through the production processes and distribution networks so they could operate at close to full employment (Chandler 1990). From 1930 to 1960, the majority of advanced capitalist countries were dominated by Fordism[1]. Ford ultimately made everything needed for its cars in-house, from raw materials to the finished product. Ford was vertically integrated for two reasons:

First, it had perfected mass production techniques and could achieve substantial economies of scale by doing everything in-house. Second, given the information-processing capabilities of the time, in addition to Henry Ford's scepticism towards accounting and finance, direct supervision could more efficiently coordinate the flow of raw materials and components through the production process than would be possible in arm's-length relationships (Chandler 1977).

However, vertically integrated production sites did have a wide range of production needs and challenges. The continuing growth of industries and the number of suppliers within them caused many firms that were vertically integrated in the early 1900s to disintegrate and form relationships with key suppliers (Chandler 1990). Some corporations began to disintegrate and outsource part of their activities to outsiders. Vertical disintegration occurred in several industries previously regarded as highly integrated, such as the automobile sector. This led to the emergence of new intermediate markets which divided a previously integrated production process between two sets of specialized firms in the same industry. Standardized information, simplified coordination and the introduction of new intermediate markets were some of the reasons for disintegrating the structure of the value chain, thus allowing new types of specialized firms to participate in an industry and thereby changing the industry's competitive landscape.

Vertical disintegration has had profound implications. It has changed the nature of the firms that can and do participate in a given industry. For example, in the oil industry, vertically disintegrated structures enabled engineering firms with no production capabilities or facilities to become significant players. The nature of capabilities, the types of entrants into an industry and the structure of competition can all change when disintegration happens. Jacobides (2005) uses an analogy drawn from the field of biology: 'Vertical disintegration allows a new, vertically co-specialized ecosystem to compete and cooperate with the old, integrated structures, thus altering the landscape for all involved'. The trend of vertical disintegration during the global business revolution caused a narrowing of the range of business activity undertaken by the individual firm. As a result of the disintegration process, a massive restructuring of assets occurred, with firms extensively selling off 'noncore businesses' in order to develop their 'core businesses' (Nolan et al. 2002).

Outsourcing

The disintegration of vertically integrated firms compelled the majority of firms to face the decision of whether to either keep key production and service activities in-house or to purchase these goods and services from other companies. Because few firms would consider managing an infinite number of activities associated with the production of output, most firms opt to source intermediate products and services from external vendors.

While managing a particular stage of production in-house may give firms the assurance of always having the optimum number of outputs, there are fixed costs associated with each stage of production. From time to time, installed in-house capacity may exceed or fall short of what is needed while fixed costs remain constant. The more activities the firm itself undertakes, the higher its investment in equipment and staffing levels will be; these costs cannot immediately be scaled up or down in response to changes in demand. Flexibility for the firm decreases, while risk increases considerably. Provided that one of the production stages is associated with a vendible good or service, a firm may endeavour to deal with surpluses or shortages by selling directly to the market. To do this effectively, however, a firm must set up and meet the costs of a marketing and buying department that will only operate occasionally as required. Alternatively, the firm may choose to abandon planned in-house production in favour of purchasing through markets and under market coordination principles. The value of resources that firms put at risk is considerably lessened when intermediate goods that serve as substitutes in production are readily available. Interestingly, this reduction in risk for the firm is not offset by an increase in the risk borne by its suppliers, as suppliers benefit from the pooling of resources (Richardson 2000).

The delivery of a continuous volume of supply does not require ownership by a single supplier. Indeed, a high volume of reliable suppliers allows firms to obtain supplies on schedule and to the right specification through enforceable contracts (Chandler 1990a). The practicality of using external suppliers depends on the availability of appropriate intermediate products and services the firm can substitute for stages in the in-house production process. As inputs and intermediary services are often provided by many different suppliers, the recipient firm must ensure careful coordination between these parties to make certain goods are produced to the specifications required.

The firm producing the final product and other suppliers engaged in the value chain have particular functions directly relating to the knowledge and expertise they possess. Not only do they require professional knowledge and technical skill, but they also must possess knowledge of each other—in other words the capabilities, roles and rules of each component of the supply chain. This creates intense interdependence and stability over the years. By working for a particular firm for a period of time, a supplier is able to provide a service which, by virtue of the supplier's experience in the industry and with the firm, becomes specialized. Therefore, the services of this supplier are no longer directly equivalent to other services available on the open market. The value of this specialization differs according to the function concerned (Richardson 1998).

A primary example of supplier specialization is found in the aircraft industry. Modern aircraft and engines are complex products. A major capability of a firm producing aircraft is the ability to integrate the whole system of suppliers to produce the final product. The core firm (the firm shaping the final product), or the supply chain system integrator, increasingly focuses on coordinating and planning within the supply chain, rather than direct manufacturing. In this industry, as much as 60 to 80% of the value of the end product (the aircraft) is delivered by the external supply network (Nolan et al. 2007). These products are highly specialized and cannot be bought in the open market because previous coordination expertise and a degree of stability in the buyer–supplier relationships is key to the ability of the firm to deliver the final product.

A high degree of stability is required in the relationship between suppliers of intermediate goods and the core firm. This relationship must remain flexible as suppliers and buyers may change and organizations need to adapt to these new relationships. However, without some degree of stability, organized coordination would not work (Richardson 1998).

Coordination by the Core Firm: The System Integrator

The requirement for stability has changed the scope and nature of the coordination function of the firm while expanding the legal boundaries of the firm. After vertical disintegration, a firm's economic activity has expanded to the coordination of activities of multiple suppliers and service providers, all of which provide input at various stages of production. The result of this coordination is the final product manufactured by the core firm.

The new structure of economic activity requires synchronized coordination of resources and production factors. Firms exist because they enable actions to be carried out concurrently in conformity with a particular design. The concurrent coordination is not an unintended consequence of market transactions; it requires organization and planning, an activity that firms typically provide (Richardson 1998). As previously mentioned, the coordination function of the firm not only encompasses coordination of resources within a firm but also coordination of the activities of other firms. Therefore, firms create a new form of industrial planning in an economy (Nolan et al. 2002).

According to Richardson (1972, 885), the orchestration of development, manufacturing and marketing occurs between retail companies and their suppliers without the existence of any shareholdings or long-term contracts:

> Co-operation takes place between firms that rely on each other for manufacturing or marketing and its fullest manifestation is perhaps to be found in the operations of companies such as Marks & Spencer and British Home Stores. Nominally, these firms would be classified as retail chains, but in reality they are the engineers or architects of complex and extended patterns of co-ordinated activity. Not only do Marks & Spencer tell their suppliers how much they wish to buy from them, and thus promote a quantitative adjustment of supply to demand, they concern themselves equally with the specification and development of both

> processes and products. They decide, for example, the design of a garment, specify the cloth to be used and control the processes even to laying down the types of needles to be used in knitting and sewing. In the same way they co-operate with Ranks and Spillers in order to work out the best kind of flour for their cakes and do not neglect to specify the number of cherries and walnuts to go into them.

Marks & Spencer plan and coordinate economic activities that do not fall directly under their remit as a firm, and as a result many economic decisions are taken due to Marks & Spencer's management and not due to 'markets', 'the price mechanism' or 'the invisible hand' (Richardson 1998).

Firms coordinate their activities by means of a 'dense network of cooperation and affiliation'. Contracts between firms are created through a balance of rewards and penalties based on the interests of cooperating parties within the organization. However, coordination is not only exercised on the formal basis of contracts. A manufacturer might not have a legal contract with a firm. Yet, based on its previous relationship with the core company, the manufacturer may take the necessary investment and organizational decisions with the expectation that the core company will continue to have a good business relationship with the manufacturer. In this sense, companies coordinate their activities with a conscious design (Richardson 1998). Conscious design is a distinct management function exercised at different levels throughout the organization, both at the firm level and at the value chain level.

As large firms have 'disintegrated', the extent of conscious coordination over the surrounding value chain has increased. In a wide range of industries, organization of the value chain has developed into a comprehensively planned and coordinated activity. The core systems integrator, in other words the 'core firm' that manufactures the final product, is at the centre of conscious coordination (Nolan et al. 2002). Large global corporations (the 'core firm' at the centre of the value chain that produces the final product) coordinate, appraise and plan activities of other firms on the same value chain through system integration. Nolan et al. (2007) assert that core firms exercise tight control over other firms across the whole value chain: 'Firms that wish to be selected as "aligned" or "partner" suppliers to the leading systems integrators, must agree to cooperate with core firms within the sector in opening their books, planning their new plants, organizing their R&D, planning their production schedules and delivering their products to the core firms'. Control by the core firm therefore covers a wide range of activities, including merger and acquisition activities, R&D decisions, marketing, sales and customer relationships with firms across the supply chain (Nolan et al. 2007).

The core system integrator coordinates extremely complex value chains that are generally comprised of several layers formed by first-, second- and third-tier suppliers, distributors and retailers (Nolan 2000). Coordination by the core firm frequently takes place within the framework provided by subcontracting between the core company and its suppliers that provide intermediate goods.

Subcontracting activity is very important because it represents a significant percentage of outputs across many of the leading global economies. For example, a quarter of the output of the Swedish engineering industry is made up of subcontracted components, while for Japan the corresponding figure is about a third. In Japan's automobile industry, subcontracted components represent nearly one half of all outputs. Subcontracting on an international basis, moreover, is becoming increasingly widespread, and now a dense network of suppliers and purchasers link industries on a global level (Richardson 1998).

An analysis of the procurement budgets of large corporations shows the wide scope of their subcontracting activity. Several suppliers 'work for' the core system integrator and produce goods that will be sold to that entity. Nolan et al. (2005) estimate that a core firm with 100,000 to 200,000 employees could easily have the equivalent of 400,000 to 500,000 employees working in companies that manufacture goods or provide services to the core firm and whose work is therefore coordinated by the core firm. For example, Nike, the athletic footwear company, does not manufacture its own products but contracts its manufacturing to 600 contract factories mainly in the Asia-Pacific region. In addition to Nike's own employment of 38,000 people worldwide, approximately 800,000 workers are employed in Nike-contracted factories around the globe (Nike 2011, 2012). In Nolan's terms, the core firm, Nike, and the firms manufacturing goods for Nike together form an 'external firm'. 'External firm' is a term used to describe the coordinated business activities that surround modern global corporations (Nolan et al. 2002).

System integrators must coordinate the activities across the value chain in order to remain competitive. Coordination capability is in itself a source of competitive advantage. A large firm gains competitive advantage by interacting effectively with a wide network of businesses. The firm needs to develop strong business relationships and local know-how with the suppliers that sell material inputs necessary for the final product. It also needs to build local knowledge relating to organization, capabilities and limits, in addition to understanding the rules and cultural requirements of supplier firms. According to Richardson (1998), coordination of the value chain cannot be obtained by telling each company what to do but must be achieved by ensuring an optimal combination in accordance with the prescribed roles and capabilities of each firm on the supply chain. Nolan et al. (2007), however, argue that in some respects, the core firm tells its suppliers exactly what to do and how to produce. In both arguments, the competitiveness of the large firm depends not only on professional knowledge and technical skills but also on the local knowledge gained from working with other firms on the supply chain. The effectiveness of the firm depends on its ability to organize the supply chain as well as on the quality of its subcontractors.

In other words, competitive advantage for successful, globalized core firms is maintained by ensuring they are at the centre of a global network of effective, cost-minimizing businesses whose interests are in close alignment with the interests of the core firm. The core firm provides direction to its suppliers so that it has access to high-quality and low-cost goods and services, which are necessary inputs to meet the end consumer's wants and needs (Nolan 2000).

The extensive coordination of the supply chain has several advantages for subcontractors. It brings stability to the business relationship between different suppliers and the core firm. It allows the subcontractor to assume targeted risks inherent in specialized skills and equipment. The subcontractor benefits from economies of scale and specific expertise achieved over years of supplying specific goods and services to a core firm. It also enables the development of specifications, processes and designs of intermediate products and services in line with the needs of the final product or service. As a result, subcontractors are able to better understand the needs of their customers and develop technologies and products accordingly.

The detailed management of the supply chain by the core system integrator results in key changes regarding the business relationship between the supplier and the core firm. Leading first-tier suppliers establish a long-term 'partner' or 'aligned supplier' relationship with core system integrators. They plan the location of their plants in relation to the location of the core system integrator. In some cases, an aligned supplier produces goods within the facilities of the system integrator. It is common practice for the leading suppliers of specialist services, such as data systems, to physically work within the premises of the systems integrator. They also plan their R&D in close consultation with the projected needs of the core firm and coordinate their product development accordingly. Production and supply schedules are comprehensively coordinated to ensure that required inputs arrive exactly when they are needed. The relationship between the suppliers and the core firm is therefore far from being a simple price relationship: the system integrators deeply penetrate the value chain both upstream and downstream, becoming closely involved in business activities that range from long-term planning to meticulous control of day-to-day production and delivery schedules, all with the aim of developing competitive advantage (Nolan et al. 2005).

It should be borne in mind that highlighting the cooperative element in business relationships and the management of the supply chain by the core firm does not mean that there is no more competition. Marks & Spencer can drop a supplier; a subcontractor can seek another principal; technical agreements have a stated term and conditions on which they may be renegotiated; the licensee of today may become the competitor of tomorrow. In these circumstances, competition is still at work even if it has changed its mode of operation (Richardson 1972).

Value Chain Management by the Core Firm

The vigorous coordination of economic activity by the firm across the value chain opens up new avenues in the debate regarding the functions of the firm. The view that a firm is a single cell in a larger organism, mainly unconscious of the wider role it fills (Dobb 1925), is now challenged by the argument that firms play an important role in the control, coordination and allocation of economic activities. As Nolan (2000) highlights, the core firm is a hub for complex planning, linking together the activities of a wide range of other firms in the pursuit of a common aim.

The scope of influence of the firm has widely expanded while the business environment that the firm controls has also grown. To gain and sustain a competitive advantage, a firm needs to formulate its strategy in an effective way to cope with and, if possible, to influence its changing environment (Kimura 2006). Large core firms need to consider the interests, costs and capabilities of the whole value chain in order to remain competitive. The core firm needs to manage the whole value chain by identifying the strengths and weaknesses of each of the other firms within the chain.

If coordination by the core firm is envisaged in terms of matching quantity and specification of intermediate output activities with final output activity, then it can be concluded that the influential scope of the firm is growing. The core firm producing or marketing the final product performs a variety of activities in the field of product development, product specification and process control that may be beyond the capability of the supplier firms. There is extensive transfer, exchange or pooling of technology through inter-firm cooperation. Thus a subcontractor commonly complements his own capabilities with assistance and advice from the firm he supplies to (Richardson 1972). In addition, firms minimize waste in competence utilization by coordinating their own competencies with the competencies of other firms to carry out complementary activities. Moreover, by consolidating its competence with that of the other firms, the firm challenges the domain of activities it is unable to do alone (Kimura 2006). The growing influence of the core firm blurs its structural boundaries and challenges the relationships it maintains with supplier firms.

The coordination function of the firm in managing its own internal resources and capabilities is extended to a wider range of planning that encompasses the resources and capabilities of other firms across the value chain. Through increased planning by systems integrators facilitated by recent developments in IT, the boundaries of large corporations have become significantly blurred. The core system integrators operating across a wide range of sectors have become the coordinators of a vast array of business activity outside the boundaries of a formal legal entity in terms of ownership (Nolan et al. 2002). Functions of the core systems integrator have changed radically away from direct manufacturing towards 'brain' functions of planning and coordination across the value chain (Nolan et al. 2002). There is a new form of industrial planning by the system integrators, and this new form extends across the boundaries of formal ownership structures and radically undermines the old size and nature of the firm (Nolan et al. 2007).

In summary, the value chain is increasingly 'managed' by the core firm. This blurs the boundaries of the firm regarding what activities fall under the firm's remit. As stated previously, the core firm and its suppliers form what is often referred to as the 'external firm'. The external firm, composed of a myriad of large, medium and small companies, operating across the value chain, increasingly integrated with information technology, works very closely in a coordinated fashion under the global management of the core firm (Nolan 2000). The separation of ownership and control in the modern corporation as argued by Berle and Means (1967) would indicate in this case that the management of the core firm not only oversees and controls the activities of the core firm for its shareholders, but it also 'manages and controls' the activities of all other firms involved in the value chain on behalf of their shareholders. This brings a different perspective to the debate on the divergence of interest between ownership and control and could be a subject in itself for further debate.

NOTE

1. Fordism, named after Henry Ford, the founder of Ford Motors, refers to the management system in operation during those decades, where large, vertically integrated global firms were producing standardized products for mass consumption.

REFERENCES

Berle, A., and G. Means. 1967. *The Modern Corporation and Private Property*. New York: Harcourt, Brace & World.

Chandler, A.D. 1977. *The Visible Hand: The Managerial Revolution in American Business*. Cambridge: Belknap.

Chandler, A.D. 1990. "Response to the Contributors to the Review Colloquium on 'Scale and Scope'." *Business History Review* 64: 736–58.

Chandler, A.D. 1992. "Organisational Capabilities and the Economic History of the Industrial Enterprise." *Journal of Economic Perspectives* 6: 79–100.

Coase, R.H. 1937. "The Nature of the Firm." *Economica* 4: 386–405.

Coase, R.H. 1990. "Accounting and the Theory of the Firm." *Journal of Accounting and Economics* 12: 3–13.

Coase, R.H. 1992. "The Institutional Structure of Production." *American Economic Review* 82: 713–19.

Dobb, M. 1925. *Capitalist Enterprise and Social Progress*. London: Routledge.

Jacobides, M.G. 2005. "Industry Change Through Vertical Disintegration: How and Why Market Emerged in Mortgage Banking." *Academy of Management Journal* 48: 465–98.

Kimura, S. 2006. "Co-Evolution of Firm Strategies and Institutional Setting in Firm-based Late Industrialisation: The Case of Japanese Commercial Aircraft Industry." *Evolutionary and Institutional Economics Review* 3: 109–35.

Langlois, R.N., and N.J. Foss. 1997. "Capabilities and Governance: The Rebirth of Production in the Theory of Economic Organization." *KYKLOS* 52. Available at SSRN.

Langlois, R.N., and R.G. Koppl. 1991. "Fritz Machlup and Marginalism: A Re-evaluation." *Methodus* 3: 86–102.

Loasby, B. 1976. *Choice, Complexity, and Ignorance*. Cambridge: Cambridge University Press.

Machlup, F. 1967. "Theories of the Firm: Marginalist, Behavioral, Managerial." *American Economic Review* 57: 1–33.

Nike. 2011. "Annual Report: Form 10-K." Accessed 26 November 2012. http://investors.nikeinc.com/files/doc_financials/AnnualReports/2011/docs/Nike_2011_10-K.pdf.

Nike. 2012. "MAS: Working Conditions." Accessed 26 November 2012. http://www.nikebiz.com/crreport/content/workers-and-factories/3–1–0-overview.php?cat=overview.

Nolan, P. 2000. "Global Business, Value Chains and Developing Countries." ECR Academic Report.

Nolan, P., D. Sutherland, and J. Zhang. 2002. "The Challenge of the Global Business Revolution." *Contributions to Political Economy* 21: 91–110.

Nolan, P., J. Zhang, and C. Liu. 2005. "The Global Business Revolution, the Cascade Effect and the Challenge for Catch-up at the Firm Level in China." Globalisation and International Business, MBA Module.

Nolan, P., J. Zhang, and C. Liu. 2007. *The Global Business Revolution and the Cascade Effect*. New York: Palgrave Macmillan.

Penrose, E.T. 1959. *The Theory of the Growth of the Firm*. Oxford: Oxford University Press.

Richardson, G.B. 1972. "The Organisation of Industry." *The Economic Journal* 82: 883–96.

Richardson, G.B. 1998. "Production, Planning and Prices." DRUID Working Paper. Danish Research Unit for Industrial Dynamics.

Richardson, G.B. 2000. "The Organisation of Industry Re-visited." DRUID Working Paper. Danish Research Unit for Industrial Dynamics.

Williamson, O.E. 2002. "The Theory of the Firm as Governance Structure: From Choice to Contract." *Journal of Economic Perspectives* 16: 171–95.

2 Petroleum Value Chain & Its Transformation

The oil industry value chain shows patterns similar to those of other sectors. IOCs act as 'core system integrators' and manage the value chain to a great extent. However, due to its specific particularities, the oil industry also shows differences to other sectors. Chapter 2 will provide an overview of the petroleum industry, its supply chain, its particularities and differences and its transformation.

2.1 PETROLEUM INDUSTRY VALUE CHAIN

Due to its geostrategic, political and economic importance, the oil industry has become one of the most prominent industries in the world. Most countries are heavily dependent on oil as a high proportion of industrial and economic activity is fuelled by oil production and consumption.

Extracting and producing crude oil and transforming it into a useable product such as gasoline or diesel is challenging. Petroleum crude oil is a natural resource that is found in tiny pores of sedimentary rocks. Typically, such rocks are found both underground and under the sea. Once extracted, crude oil must be refined in order to be made ready for consumption. Due to the difficulty of extracting crude oil, the often-remote location of oil fields and the sophisticated oil-refining processes, the oil supply chain is extremely complex. This supply chain can be divided into two major segments: the upstream and downstream supply chains. The upstream supply chain involves the acquisition of crude oil and consists of exploration, forecasting, production and logistical management of delivering crude oil from remotely located oil wells to refineries. The downstream supply chain starts at the refinery where crude oil is processed into consumable products.

The downstream supply chain involves forecasting demand, production and managing the logistics of delivering oil products to customers around the globe. Most oil companies concentrate their activities in the upstream supply chain, whereas refineries and petrochemical companies concentrate on the downstream segment (Hussain et al. 2006). Integrated oil companies, however, are involved in both the upstream and the downstream supply chain. Both ends of the supply chain consist of multiple activities involving a wide range of companies and stakeholders.

Upstream supply chain activities involve a range of types of companies. Before exploration work starts in an unexplored basin, a legal framework must be in place so that oil companies are guaranteed a legal right to extract any oil deposits they discover. Host governments auction leases for exploration acreage, and in some cases they commission seismic surveys of the acreage under offer to provide basic information to prospective bidders. Once acreage is obtained,[1] an oil company will commission a seismic survey from which potential reservoir targets are assessed. A seismic survey requires seismic companies, seismic data brokers, geologists and geophysicists. Once several sites have been investigated and ranked in order of attractiveness, a drilling company and associated service companies (supply boats, helicopters, cementing, mud logging, artificial lift, coil tubing companies, etc.) are hired and the target is drilled up. With skill and luck (even with the advantage of modern seismic technology, the chance of finding oil or gas is still less than 20%), the oil company uncovers crude oil deposits during a drilling campaign. Once a discovery has been made, the oil field must be developed—a challenge which requires considerable time.

As an example, development of the Northern North Sea oil field took twelve years between the peak in discoveries and the peak in production. Oil field development involves drilling production wells (and if need be, injection wells) and building infrastructure such as platforms, pipelines, processing plants and export terminals. This requires huge capital expenditure outlays and involves a large number of oil services companies, ranging from engineering and construction companies to offshore rig builders. Following development, the oil field experiences two phases of production: a peak phase and a declining phase. During the declining phase, it becomes increasingly more difficult to extract crude oil from the field. This causes the field's free cash flow to diminish as lower volumes of oil are extracted at a higher cost. In some cases, these fields become uneconomical for large oil companies due to their high cost structure. They then dispose of declining oil fields by selling them to smaller oil companies, such as Venture, Tullow, Dana or Apache. Smaller companies take over declining fields and extend their useful life by using enhanced recovery techniques. When the cost of extracting the remaining oil is not justified by oil prices, the fields are abandoned. For onshore sites, the oil wells need to be plugged with cement or steel plugs; for offshore sites, large platforms must be dismantled using large cranes. The disposal of oil fields requires a high capital expenditure and is quite risky. Companies specializing in decommissioning and dismantling are contracted to assist with this work (Hermann et al. 2010). As an example, Shell has recently announced the decommissioning and dismantling of its installations in the Brent Field.[2] Shell estimated that cleaning up the whole field, which has four platforms, will require a decade and cost billions of pounds (NY Times 2015).

Table 2.1 Upstream Activities, Services and Equipment at the Oil Field

Identify Targets	Drill Exploration & Appraisal Wells	Develop the Field	Extend the Field Life	Decommission the Field
Seismic	Drilling services	Drilling	Infill drilling	Diving
Geologists	Associated well head services (well preparation, pressure pumping, directional drilling, completion)	Data acquisition	Injection wells	Heavy lift equipment
Geophysicists	Data acquisition (mud logs, well site geologists, coring, wireline logs, logging while drilling, well test)	Engineering & construction services	Gas lift	Large crane vessels
		Platforms	Pumps	
		SURF	Fracturing	
		Pipelines, rigid and flexible pipes		
		Topsides and hulls of platforms		

Source: Author's compilation based on the figure 'The life cycle of an oil field' (Hermann et al. 2010, 50).

The downstream supply chain, on the other hand, starts once oil is extracted and appraised. At that point, processing and commercialization of the commodity can begin. Once the crude oil is pumped out, it is transported from production facilities to refineries by crude oil tankers, tanker trucks, railways or pipelines.

Crude oil is then transformed into a variety of petroleum products inside a refinery. The basic refining processes are separation, conversion, purification and blending.[3] Suppliers of equipment (e.g., hydrocrackers) or chemicals used during the refining process provide goods and services to the refining company and hence play a critical role in the downstream supply chain.

Once the finished petroleum products are ready, they are transported to markets mostly via pipelines, oil tankers, tanker trucks and railroad tank cars.[4] At local terminals, distributors collect oil products and deliver them to underground tanks at service stations, convenience stores, petrochemical companies and other retail outlets. In addition to the oil refineries and pipelines, a network of independent sea-fed terminals transport oil products to distributors. These terminals are typically owned by specialist storage companies that rent their tank capacity to other companies. Regardless of its origin and the method by which it is produced and distributed, almost all oil is delivered to service stations by large tanker trucks. A significant share of the oil is transported by oil tankers, making shipping an important segment in the downstream value chain. Vessels are built in shipyards and are owned by and registered in the name of ship owners. Ship owners may delegate the management of the ship to a third party. In this case, the management company deals with the technical and nautical handling of the vessel. It takes out insurance, recruits the crew, checks the certification documents, carries out maintenance and inspects the ship's condition. Each vessel is classified by certification agencies according to the ship's structural integrity and the reliability of its machinery and equipment. IOCs charter oil tankers from ship owners or their management companies. Last, the large refineries and natural gas processing plants provide feedstock for petrochemical companies.[5] Historically, the oil and gas companies' involvement in the petrochemical industry came from their desire to add value to certain by-products arising from the refining of crude oil and natural gas. Petrochemical companies manufacture base chemicals and convert these into basic plastics. Integrated oil companies such as Total, Shell and Exxon are among the world's top-ten chemical companies by revenue (Hermann et al. 2008).

2.1.1 Companies on the Petroleum Value Chain

The petroleum industry's value chain from upstream exploration and production to downstream refining, sales and marketing is very broad and involves diverse companies. The companies in the upstream segment, upon which this book focuses, can be grouped into four main categories, namely

Table 2.2 Downstream Activities, Services and Equipment from the Oil Field to Customer

Production Onshore & Offshore Fields	Transport from Fields to Refineries	Refining	Transport from Refineries to Marketing & Petrochemical Plants	Marketing
Systems used for production	Receipt/Loading terminals	Refining units (separation, conversion, purification, blending)	Pipelines from refineries to distribution terminals/oil storage terminals	Distribution
Platforms	Pumps		Independent sea-fed terminals	Primary terminals
Oil rigs	Crude ships		Specialist storage companies	Trucks
Pipelines	Pipelines		Petrochemical plants	Secondary terminals
Pumps	Railways		Oil tankers	
	Unloading terminal			
	Refinery			

Source: Compiled by the author.

international oil companies, national oil companies, independents and oil services companies, according to the nature of their ownership structure and the scope of their business activities.

Although a consistent definition of independent oil companies does not exist, 'the independents' are known as nonintegrated companies which are focused on the exploration and production segments of the oil and gas industry and which are not involved in marketing or refining. These are mainly small companies involved in the development of domestic oil and gas businesses, although there are also larger ones engaged in overseas business, such as Anadarko.

National oil companies are state-owned or -controlled companies whose role in the world oil economy has changed dramatically in recent years. While in the 1970s the IOCs had 85% of the world's hydrocarbon resources on their books, the present period is marked by the surge of NOCs as leading producers of oil and gas following a series of nationalizations during the '70s, as shown in Figure 2.1.

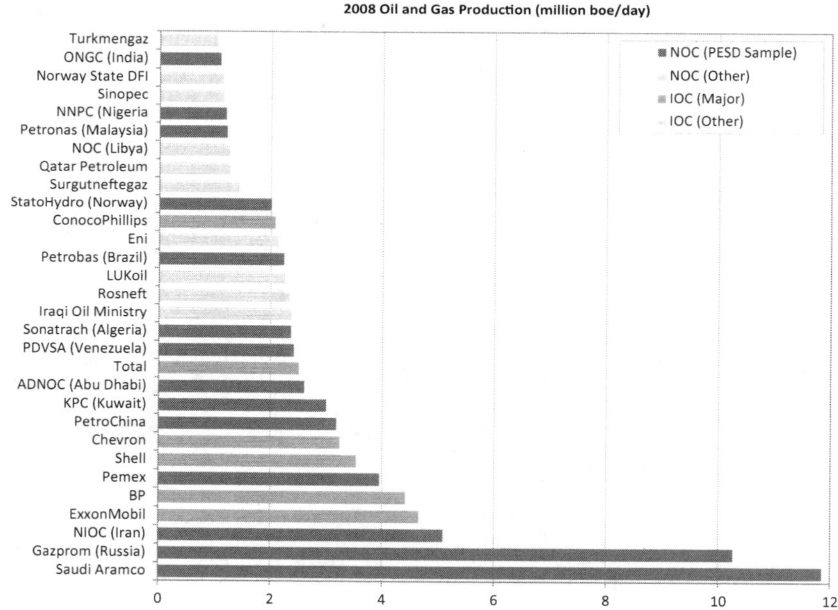

Figure 2.1 NOCs as Leading Producers of Oil and Gas

Source: Figure based on the figure in Thurber (2012, 3), based on source data from Wood Mackenzie.

Note: PESD is the Programme on Energy and Sustainable Development at Stratford University, where Thurber's data analysis took place.

Estimations regarding the oil reserves controlled by NOCs and IOCs vary but are in agreement that NOCs hold the lion's share. Some sources state that NOCs control 80% of world oil reserves and 73% of oil production, while other observers place them even higher, at 92 to 95% of oil reserves (Howat 2006; James 2011; Jessen 2009; Victor 2007; Yergin and Stanislaw 2002).[6]

With their vast control of oil production, NOCs form thirty out of the top fifty largest oil companies according to *Petroleum Intelligence Weekly's* 2011 rankings[7] (Broxson 2012). Furthermore, based on oil and gas reserve holdings, NOCs represent the top ten largest companies while ExxonMobil, the biggest IOC, ranks as fourteenth and Shell as twenty-fifth (Jaffe and Soligo 2007). NOCs are therefore the largest players in the oil industry. They are however far from monolithic. Some NOCs are 100% state controlled, such as Saudi Aramco and NIOC, whilst others have only a minority state ownership, such as Petrobras (31%) or ENI (30.3%). According to industry experts, some NOCs are at the far front of leading edge technology, such as Saudi Aramco, Petrobras or Statoil, while others, such as Sonatrach or INOC, fall short of possessing similar technological expertise. Most NOCs have mixed missions ('national' and 'commercial') and thus do not always perform as profit-maximizing enterprises. They tend to develop their reserves to a lesser extent than IOCs and not generate revenue from their output with the same efficiency (Victor 2007). However, they are important players that impact the relationship between IOCs and OSCs.

IOCs, on the other hand, present more consistent characteristics. They are privately owned companies that aim to maximize profit for their shareholders through oil and gas extraction and monetization on a global scale (James 2011). They are non-state owned, vertically integrated oil and gas companies involved in all stages of the oil industry, including exploration, production, refining, trading, marketing and sometimes transportation. They are integrated through upstream to downstream. There is a lot of debate regarding the rise of NOCs and the competitive pressure that this causes on IOCs. In fact, IOCs and NOCs characteristically supply very different functions to governments when it comes to managing risk in oil exploration and exploitation (Thurber and Nolan 2010).

Finally, oil services companies are specialized companies that provide technology and equipment to IOCs and NOCs. They often occupy a particular niche that involves providing a specific technology, equipment or service to their customers. A few larger OSCs provide a wide range of specific technologies.

Because the main aim of this book is to investigate the transformation of the oil industry via analysis of the relationship between IOCs and OSCs, particular attention will now be given to these entities.

International oil companies are the large integrated oil and gas companies. The oil industry consolidation, an outcome of the search for competitive advantage, has created a set of companies known as the 'oil majors'

(OMs) or 'super majors'. From 1998 to 2000, the number of Western oil companies reduced from eleven to six, and the giant corporations ExxonMobil, BP, Total, Chevron, Conoco Philips and Shell were created (Zhang 2004). These six large and dominant firms now rank among the world's biggest companies in terms of revenues, profits and market value (Forbes 2011; FT 2011).

In line with the changing nature of the firm and due to the increasing coordination function of core system integrators in global business, IOCs work as core system integrators in the oil business and organize the value chain to a great extent. They possess key attributes, including the capability to finance large projects and the resources necessary for R&D alongside the ability to develop a global brand and attract a high-calibre work force (Nolan et al. 2002). They offer risk management functions to host countries by taking risks, such as those related to geological uncertainty and future market conditions. Moreover, they invest capital in an unknown reserve and create a bridge to sources of demand for oil products (Thurber and Nolan 2010).

The five main companies which will be analysed under the IOC rubric within the context of this book are ExxonMobil, Shell, BP, Chevron and Total. These companies, which are aptly named oil majors, are the biggest IOCs by market cap and revenue. ConocoPhillips, which is also considered as an OM, will not be included in the analysis because it has recently changed its business model from that of an integrated company to a pure exploration and production company.

The majority of IOCs gained their 'oil major' status as a result of several mergers and acquisitions. Except Shell and Total, they can all trace their

Table 2.3 Major Oil Companies by Market Cap

Company	Headquarters	Market Cap (Million USD)	Revenue—T12M (Million USD)
ExxonMobil	US	364,634	329,647
PetroChina	China	356,197	370,538
Shell	Netherlands–UK	198,735	377,153
Chevron	US	198,372	173,714
SINOPEC	China	140,870	458,685
BP	UK	131,083	316,054
Total	France	128,652	194,126
ConocoPhillips	US	80,857	44,825
Gazprom	Russia	70,343	147,071
ENI	Italy	68,611	133,067

Source: Compiled by the author with Bloomberg data (as of May 22, 2015). Used with permission of Bloomberg L.P. Copyright© 2015. All rights reserved.

Table 2.4 Selected Financial Information of International Oil Companies

	ExxonMobil		Shell		BP		Total		Chevron	
(BN$)	**2014**	**2013**	**2014**	**2013**	**2014**	**2013**	**2014**	**2013**	**2014**	**2013**
Sales	412	438	431	460	359	396	236	251	212	229
Purchases	–226	–244	–327	–353	–282	–298	–152	–160	–120	–135
Production and Manufacturing Expenses	–41	–41	–30	–28	–27	–28	–28	–28	–25	–25
Selling, Distribution, General and Administrative Expenses	–13	–13	–14	–15	–13	–13	n/a	n/a	–5	–5
Exploration Costs	–1.7	–2.0	–4.2	–5.3	–3.6	–3.4	–1.9	–2	–2.0	–1.9
R&D Expenses	–1.0	–1.0	–1.2	–1.3	–0.7	–0.7	–1.3	–1.2	–0.7	–0.8
Net Income	33	33	15	16	4	23	4	11	19	21
Capital Expenditure	–39	–42	–32	–40	–23	–25	–30	–34	–40	–42

Source: Compiled by the author based on annual reports of each company.

Note: All figures are billion USD. All sales figures include revenues and other income. TOTAL reports its production and administrative expenses together as operating expenses.

history back to Standard Oil. Standard Oil was formed by John D. Rockefeller in 1870 and became the first and biggest integrated oil company of its time, producing, transporting, refining and marketing oil in the US (Yergin 1991). It was found guilty of monopolizing the petroleum industry in 1911[8] and dissolved into thirty-four smaller regional companies as a result of a court order. These companies[9] formed the basis of the current ExxonMobil, BP, Chevron and Conoco.

ExxonMobil: Today's Exxon stems from four Standard Oil companies,[10] one of which was Standard Oil New Jersey, the heart of Standard Oil. Its current form took shape in 1999 with the merger of Exxon and Mobil, the two largest descendants of Standard Oil. The historic $81 billion deal was the largest corporate deal in US history. The merger, the largest in a string of consolidating moves within the industry, was defended by Exxon citing price pressure on crude oil, the need for greater efficiency and new competitive threats overseas (CNN 1999). 'This merger will enhance our ability to be an effective global competitor in a volatile world economy and in an industry that is more and more competitive', said Lee Raymond and Lou Noto, chairmen and chief executive officers of Exxon and Mobil, respectively. ExxonMobil has since become the leading IOC. It is the world's largest publicly traded international oil and gas company with the highest inventory of oil and gas resources among IOCs. Its reserves amount to 87 billion net oil-equivalent barrels as of 2011, with a production of 4.5 million barrels per day on an oil-equivalent basis. It is an integrated company owning the biggest refinery network and is one of the largest chemical companies in the world. In 2011, it generated a net income of $41 billion and undertook capital and exploration expenditures of $37 billion (ExxonMobil 2011, 2012).

Chevron: Chevron has its origins in Standard Oil California. Its dramatic discoveries in Saudi Arabia have helped the company to become a major oil company. Standard Oil of California (later Chevron) began exploring in the kingdom's Eastern Province in the early 1930s and formed California Arabian Standard Oil Company (CASOC). The Texas Co. (later Texaco) joined in 1936 as a partner to CASOC, which became the Arabian American Oil Company, or Aramco[11] in 1944. CASOC discovered oil in Dammam Dome No. 7 in 1938. Continued exploration in the same area led to several discoveries including, in 1948, Ghawar—the world's largest oil field in terms of production and reserves. Access to reserves in Saudi Arabia provided Chevron with low-price, high-grade oil and ensured one of the best profit ratios among all oil companies during the 1950s and 1960s. Chevron lost its favourable position in Saudi Arabia during the period of nationalization. It merged with Gulf in 1984, the biggest merger ever at that time, and with Texaco in 2001 to create today's Chevron. Chevron is the second-largest integrated company in the US, producing 2.6 million barrels of oil-equivalent per day. In 2011, Chevron generated a net income of

$27 billion and spent $29 billion on capital and exploration expenditures (Chevron 2011, 2012; Funding Universe 2012b).

Shell: Royal Dutch Shell Group was formed in 1902 by the merger of British Shell Transport (created as an oil shipping company in 1878) and Holland's Royal Dutch (created in 1890 to explore oil in the Dutch East Indies). In 2005, as a result of major structural reorganization, the hundred-year partnership was dissolved and a single company, Royal Dutch Shell, was created. RD Shell wasn't involved in mega mergers like other IOCs. It did, however, purchase Enterprise Oil (the UK's largest exploration and production company at that time) and Pennzoil-Quaker State (a motor oil company descendent of Standard Oil) (Hermann et al. 2010). Shell is now the second-largest international oil and gas company in terms of market capitalization, operating cash flow and oil and gas production. It produces 3.2 million barrels of gas (48%) and oil (52%) per day in 80 countries. In 2011, it generated $31 billion income and invested $26 billion capital (Shell 2011, 2012).

BP: BP has its origins in Standard Oil Indiana (Amaco) and Anglo Persian Oil. Anglo Persian Oil Company (which became Anglo Iranian in 1935 and then British Petroleum in 1954) was formed in 1908 to explore oil in Iran. From that first uncertain search for oil in Persia, BP has grown over the years to become a global energy company. In 1953, the Iranian nationalization deprived the company of two-thirds of its production. The company responded by increasing output in Iraq and Kuwait, by building new refineries and by launching further exploration activities in the Arabian Gulf, Canada, Europe, North Africa, East Africa and Australia. The company's future was secured in the late 1960s by major oil discoveries in Alaska and the North Sea. After looking for oil in the UK for fifty years without a single large discovery, British Petroleum was the first company to find oil in the North Sea. A far bigger discovery happened in Alaska, where after a decade of drilling dry wells along the North Slope, British Petroleum was on the verge of abandoning its search. Following a suspiciously extravagant offer by its competitors for its acreage, British Petroleum returned to the search and in 1969 tapped into its share of the largest oil reservoirs ever found on the North American continent. British Petroleum has been involved in several mergers and acquisitions in the past fifteen years, such as the merger with Amaco in 1998 and the acquisition of Arco and Castrol in 2000. In 2001, British Petroleum was rebranded to become BP. In 2003, BP established TNK-BP in Russia with a 50% share. These major oil findings and mergers and acquisitions (M&A) helped the company to reach its current size. After making its first annual loss in its history ($3.7 billion) in 2010 following the Macondo accident (which will be analysed in detail in chapter 3), BP made a profit of $26 billion in 2011 with production of 2.3 million barrels of oil per day (BP 2011, 2012; Funding Universe 2012a).

Total: Compagnie Française de Pétrole (today's Total) was founded by the French government to create an instrument capable of implementing a national oil policy. During the execution of the San Remo Treaty, signed in 1920, CFP inherited a share of Turkish Petroleum Company, which had been seized as war reparations. The settlement gave the company access to oil fields in Iraq. Total has become the first foreign oil company allowed back in Iran since the overthrow of the Shah in 1979 and chose to stick with its projects there despite US sanctions and opposition. Total merged with Belgium's PetroFina in 1998 and its French rival Elf in 1999. It is the fifth-largest publicly traded integrated international oil and gas company in the world, with a production of 2.34 million barrels of oil-equivalent per day in 2011. During the same period, it generated $16 billion in net income and spent $34.1 billion on capital expenditure. Total's net investment amounts to $22 billion (Total 2011, 2012b, 2012c).

ConocoPhillips: ConocoPhillips is relatively smaller in size compared to other large IOCs and is therefore sometimes excluded from the group of OMs. It is still the third-largest American oil company that can trace its history back to Standard Oil, via Continental Oil. It was created by the merger of Conoco and Phillips in 2002. As part of its strategy to reposition the company, ConocoPhillips separated the company's upstream and downstream businesses into two companies. In 2012, ConocoPhillips implemented a spinoff of its downstream business (refining and marketing, chemicals businesses) into a new company, Phillips 66, and has decided to continue its operations as an exploration and production company. ConocoPhillips has justified the move on the grounds of its intention to optimize its portfolio, enhance returns and increase financial flexibility (ConocoPhillips 2011).

Both the characteristics and the market structure of the oil industry are constantly evolving. The international oil sector has shown decades of vertical and horizontal integration. IOCs have adopted a vertically integrated structure, with their presence being felt along the entire supply chain, from exploration to refining and sale to consumers. Horizontal integration has also been implemented via the undertaking of several mergers and acquisitions within the industry. Large oil companies dominated the oil industry before the 1970s owing to their access to 85% of reserves. However, their major share in oil production came to an end with the nationalization of the bulk of the world's proven oil reserves in the late 1970s (Victor 2007). The low oil prices and low margins of the 1990s caused consolidation among private oil companies. The mega mergers in the late 1990s and early 2000s created today's IOCs. Table 2.5 summarizes the largest M&A deals among IOCs.

Even though international oil companies are among the largest companies in the world, oil market structure is dominated by NOCs, as previously stated. Following the so-called Seven Sisters[12] period where Anglo-Saxon companies controlled the majority of oil reserves in the Middle East, a profound shift in power occurred and NOCs became the new Seven Sisters

Table 2.5 Major M&A Deals in the Oil Sector (1998–2001)

Acquirer	Target	Year	Transaction Value (Billion USD)
BP	Amoco	1998	53
Total	PetroFina	1998	11
Exxon	Mobil	1999	81
BP	Arco	1999	26
TotalFina	Elf	1999	48
Chevron	Texaco	2000	35
Philips	Tosco	2001	7
Conoco	Gulf Canada	2001	6
Philips	Conoco	2001	15

Source: Author's compilation based on the figure 'Main M&A Deals in the Oil Sector' in Ochssee et al. (2010, 7).

with control of world oil reserves (Hoyos 2007). Today, large IOCs compete with independents and with medium-sized operators, such as BG Group and GDF Suez, in the global race for reserves.

OSCs, meanwhile, provide assets, equipment, technology, manpower and project management that enable oil companies, including IOCs, to explore and develop oil and gas fields. They provide a wide range of services including geophysical and geological support, drilling, well support services and engineering services for oil and gas wells. Oil equipment manufacturing companies produce oil field equipment, drilling bits, drilling fluids, drill rigs, lifts, pipes, pumps, valves, wellheads, etc.

The oil services sector involves thousands of companies which were classified into three groups by *The Economist* newspaper in 2012: companies that own and lease out drilling rigs (such as Transocean, Noble and Seadrill), companies that manufacture and sell equipment for use on drilling rigs or the seabed (including FMC, Cameron and National Oilwell Varco) and those that carry out most of the tasks involved in finding and extracting oil (dominated by the four giants Schlumberger, Halliburton, Baker Hughes and Weatherford International) (Economist 2012).

Because the aim of this book is to consider a larger scope of activities, oil services firms have been categorized into the following six major segments.

Seismic Companies: Geophysical service companies providing seismic data acquisition, processing, reservoir imaging and monitoring services. Main players are CGGVeritas, WesternGeco, PGS and BGP.

Drilling Companies: Directional drilling, land drilling and offshore drilling companies including Transocean, Ensco, Sea Drill and Nobel.

Well Servicing Companies: Technology-oriented companies in the upstream segment working on formation, evaluation, well construction and completion of oil and gas wells. High consolidation has occurred in this segment with four companies, namely Schlumberger, Halliburton, Weatherford and Baker Hughes, covering most services.

Engineering, Procurement & Construction (EPC) Companies: Companies engaged in engineering, procurement, design and construction of oil-related facilities (e.g., platforms and refineries), such as Technip, Saipem, Aker Solutions, Amec and Tecnicas Reunidas.

Subsea, SURF: Companies designing, manufacturing and servicing systems and products for subsea production and processing such as Technip, Saipem, FMC Technologies and Cameroon.

Equipment Producers and Manufacturers: Companies that manufacture equipment such as pipelines, steel pipes, valves, pumps and pressure units, including National Oilwell Varco, Dresser-Rand Group, Oceaneering International, FMC Technologies, Rolls Royce and GE.

The companies above are those that are involved directly with either the oil reserves or the equipment used in the process. There are also environmental and safety companies engaged in environmental monitoring and waste management, fabrication and construction companies (boat and rig builders, shipyards) and all other types of support companies, such as well site maintenance, helicopter, shipping, assisting boat and diving companies.

OSCs are either large companies offering a broad range of services, such as Schlumberger, or they are small specialty firms servicing a niche market, for example companies specialized in optimization of unconventional oil. Only a limited number of companies exist in certain niche markets. For example, there are only four to five companies in the subsea and deepwater well services segments. Table 2.6 lists the largest OSCs by market cap.

The expansion of the oil services sector happened after the 1970s. The oil shocks of the '70s[13] caused IOCs to downsize and outsource certain economic activities to service companies (Baxter 2009; Macalister 2011). As Andrew Gould, a past chairman and CEO of Schlumberger and now chairman of BG Group, has stated, the oil service industry was a North American affair prior to the 1970s. Outside of North America, the operators did a lot of work themselves. The outsourcing following this upheaval gave the service industry the opportunity to expand its services. 'Following the oil price collapse many operators cut back on their investment in technology and encouraged the service industry to undertake many of the tasks previously formed in house. The next step, started by the two companies (BP and Shell) in the North Sea, was to outsource whole sections of the process to the contracting industry to try to streamline costs' (Andrew Gould quoted in *Finding Petroleum* 2012).

As a result of the increased outsourcing in the oil industry, the oil services sector flourished with occasional drawbacks—the two major oil price collapses in 1986 and 1998 that changed the picture and altered its structure. In 1986, when oil prices dropped below 10 $/bbl, IOCs requested contractors to reduce their service fees. Subsequent price

Table 2.6 Major Oil Services Companies by Market Cap

Company	Headquarters	Market Cap (Million USD)	Revenue—T12M (Million USD)
Schlumberger	US–France	116,853	47,589
Halliburton	US–Dubai	39,276	32,572
Baker Hughes	US	28,636	23,414
National Oilwell Varco	US	19,936	21,371
China Oilfield Services	China	16,623	5,355
Keppel Corp	Singapore	12,061	10,185
Weatherford International	Switzerland	11,521	14,109
Offshore Oil Engineering-A	China	10,300	3,353
Cameron International	US	10,171	10,396
FMC Technologies	US	9,906	7,813
Helmerich & Payne	US	8,130	3,877
Technip	France	8,086	14,095
Transocean	US	7,300	8,878

Source: Compiled by the author with Bloomberg data (as of May 22, 2015). Used with permission of Bloomberg L.P. Copyright© 2015. All rights reserved.

Note: Bloomberg indicates 436 oil equipment and services companies in its database. Table 2.6 contains companies with a market cap larger than $7 billion as of May 22, 2015. The information regarding the company headquarters is not available from Bloomberg and has been added by the author.

discounts put OSCs in financial difficulty, forcing them to consolidate through a series of mergers and acquisitions in the '80s and '90s. According to an industry expert, the biggest and longest lasting downturn for the oil services industry began in 1982 in the US and spread to the rest of the world in 1986. The downturn lasted nearly twenty years (1986–2006), with short periods of improvement in the mid-'90s. During the downturn, numerous OSCs faced significant financial and operational difficulties. In 2004, several OSCs were on the verge of bankruptcy. For example, Halliburton posted a net loss close to $1 billion[14] each year between 2002 and 2004 (Halliburton 2004). There were industry-wide concerns that some OSCs might become insolvent. According to a few industry experts, IOCs provided support to OSCs at this time and helped them overcome these challenges.

During this period, OSCs began restructuring and investing in key technologies. Restructuring reduced the number of companies and created large OSCs. For example, the number of major seismic companies reduced from twelve to three during the downturn. Schlumberger, Baker Hughes, Halliburton and Weatherford have significantly grown into larger companies and have started offering integrated services.

The largest OSC as of December 2012 is Schlumberger. In terms of market capitalization, Schlumberger is followed by National Oilwell Varco, Halliburton, Saipem and Baker Hughes, as shown in Table 2.6. In the US, drilling represents approximately 35% of oil services industry revenue while well support services and equipment manufacturing represent 45% and 20% respectively (Hoovers 2009). The US and European service companies dominate the industry, albeit the US oil services sector has a market value almost four times greater than its European counterpart. Although there is some overlap, US service companies tend to offer drilling and well support services (e.g., mud logging, supply vessels, pressure pumping) while the Europeans focus predominantly on installation, engineering, construction and seismic activities (Hermann et al. 2008).[15]

The current global industry structure is marked by the coexistence of large and small companies and a lesser and decreasing share of middle-sized companies due to financial difficulties and technical challenges. First, financial difficulties caused by the collapse of oil prices caused consolidation in the industry, as explained above. OSCs merged due to competitive pressure during the downturn period. Second, oil projects have become larger in size and require more advanced technology, leaving few service companies able to provide the necessary expertise. Oil projects that are larger, more complex and technologically challenging can only be undertaken by financially healthy OSCs. These two occurrences, namely financial difficulties caused by the collapse of oil prices and larger and more complex projects requiring sizeable companies, resulted in large-scale consolidation in the oil services sector. The consolidation created large global OSCs that have a range of capabilities, replacing the middle-sized companies. Remaining middle-sized companies have merged in order to become bigger. For example, Acergy and Subsea 7 merged in 2010 and have started competing with large companies such as Technip and Saipem. During the merger, both companies declared that 'the new firm will be better able to meet the growing size and technical complexity of subsea projects, driven by the demand to access ever more remote reserves in increasingly harsh environments' (Goldstein 2010). Today, large OSCs either offer a wide range of services or become specialized in a particular service area. For example, Schlumberger has expanded its range of services by regularly assessing its work flows and industry requirements, whereas Cameron has specialized in providing technical equipment, particularly blowout preventers (BOPs).

The rest of the industry is characterized by numerous small companies that work for large OSCs and specialize in specific technologies—such as Reef Subsea, a small subsea contractor. These contractors work on very particular technologies, and when they grow in size, they tend to become prey for larger OSCs.

The industrial structure of oilfield services has therefore become polarized with a dominance of large or small companies gaining market share and a reduction in the number of middle-sized companies. According to an industry expert, 'Mid-size companies have been acquired by larger companies or squeezed out of the supply chain all together'. In other words, a 'cascade effect'[16] is observed mainly due to financial difficulties and technical challenges.

2.1.2 Particular Characteristics of the Petroleum Value Chain

Onshore/Offshore

Although the main activities—exploration, seismic research, drilling, etc.—are similar, the value chain differs for onshore (on land) and offshore oil fields. Offshore fields are situated miles from land and require drilling through the seabed in more difficult pressure environments. Drilling in offshore fields has to deal with the constraints of deepwater, remote logistics, absence of a base supporting the rig and often harsh weather conditions. Offshore operation environments can be very inhospitable. In the most extreme cases, pressure and temperature can reach 35,000 psi (241 MPa) and 500 °F (260 °C) in deep holes. Because offshore operations are more complicated than onshore ones, they typically involve more companies, use more advanced technology and require a greater expenditure of capital.

For OSCs involved in both onshore and offshore fields, onshore activities typically generate higher revenues, but offshore activities tend to be more profitable. For example, in 2011 Saipem received €6,685 million revenue from onshore activities and €5,908 million from offshore work. However, in terms of profitability, the offshore activities of Saipem generated more revenue than onshore, with EBITDA of €1,384 million for offshore versus €751 million for onshore. Remuneration of OSCs is often higher for offshore fields due to the nature of assets and capital that are used (Saipem 2012).

According to an interviewee, because offshore fields are more capital intensive, economic uncertainty has a greater impact on them within the oil investment cycle. While big onshore projects were awarded to OSCs during 2009–2010, the launch of new large offshore projects has decreased following the 2008 recession. The reduction in offshore projects has been caused by IOCs' preferring to delay high-risk projects.

Both onshore and offshore fields require the use of OSCs, but only a small number of companies have the necessary capability and technology for operating in harsh offshore conditions. Therefore, the supply chain is more consolidated offshore than onshore.

Advanced Technology

While both upstream and downstream are complex segments, certain sections of the upstream segment are particularly challenging from a technical point of view. For example, IOCs may need to drill up to 12,376 metres onshore[17] and have offshore production platforms that are moored in 2,450 metres of water (Russia Today 2012; Shell 2010). Some deepwater and ultra-deepwater[18] projects involve drilling in areas where waters reach depths of 1,000 to 2,000 metres, and where reservoirs are buried 1,000 to 5,000 metres below the seabed in fields located 150 to 300 km off the coast (MMS 2008). At these depths, all drilling and production work has to be done by remotely operated underwater vehicles (unmanned submarines). An advanced level of engineering is required to cope with such high pressure, high temperature (HPHT)[19] environments. Once the oil reaches the seabed from the HPHT environment, special cables are needed to protect production on its journey across thousands of metres of subsea lines, both from low temperature (4 °C) and from the extreme pressure (140 bar) prevailing at the bottom of the ocean.

In his address to the UK Parliament, Malcolm Webb, chief executive of the industry association Oil and Gas UK, presented the unique technical challenges of deepwater as greater water depths, higher pressures and the successful manipulation of the extra-long riser pipe which connects the wellhead to the rig (UK Parliament 2011). All these technical challenges are met through the use of advanced technology—such as satellite-linked dynamic positioning systems, 3D seismic imagers and remotely operated vehicles able to function at depths of up to 5 km underground—making IOCs and OSCs among the most technologically advanced companies in the world.

A good example of this sort of advanced technology is the Perdido production platform operated by Shell. It has accomplished one of the most challenging engineering feats ever achieved, producing oil and gas among the powerful hurricanes that sweep the ocean surface while extreme pressure and near-freezing temperatures pose different challenges on the rugged seabed. Tyler Priest, oil historian and professor at the University of Houston, has classified Perdido as 'the most technologically advanced facility in the world' and said that 'Perdido opens up a whole new frontier in deepwater oil production' (Tyler Priest quoted in Shell 2010). In order to deal with complexities, IOCs deploy the most advanced IT systems. In 2009, the computing power of the supercomputer at Total's Exploration & Production Scientific and Technical Centre in southwest France was increased from 122 to 450 Tflops (teraflops)—a capacity of 450,000 billion floating point operations per second. It is claimed that capacity is raised regularly and will soon reach 1,000 Tflops. The threshold of exaflops (10^{18} floating point operations/second) could be reached by 2020 (Total 2012a). Industry experts state that only space and nuclear technology use more powerful computers than those used in petroleum industry.

2.1.3 Differentiating Elements of the Petroleum Value Chain

Besides the particular characteristics of the petroleum supply chain, there are several key differences with other sectors' supply chains. In order to understand the different roles of IOCs and OSCs within the value chain, it is necessary to first look at the differentiating elements of the oil industry supply chain versus other industries.

First of all, in the oil industry supply chain, the end product is the oil, which is not a manufactured product like an automobile or aeroplane. Procurement is vital in the automotive and aerospace industries, because the goods procured are used directly as parts in the manufacturing of the car or aeroplane. In the oil industry, however, while the goods and services procured are used to build platforms and drill wells, they are not components used in the end product.

Second, the cost of the parts is essential in the manufacturing industries because they are the main components of the assembled product. The cost of OSCs in exploration and production projects is equally important as it is the primary expenditure for IOCs. However, the expenditure itself is not the main driver that determines the profitability of an oil field. The value of each field is derived from production of oil, rather than the tools that are used to reach the oil. The profitability of each field is mainly affected by the amount of oil found and by its price. Additionally, the final products in the manufacturing industry do not experience strong price fluctuations, whereas oil traded in the international market is subject to extreme price volatility. For instance, the price of Brent crude oil has fluctuated between $36 and $147 during the 2009 to 2015 period, demonstrating the high price volatility of the final product of this sector.

Third, the main characteristic of the oil industry is that the majority of the equipment and services are customized and dedicated to a specific project. While there is common equipment, such as pipes and valves, used in all field development, the majority of services and equipment are customized. Because each field development is unique, the serial manufacturing is very limited. Each oil field presents different challenges and hence requires different techniques and equipment. The technology and services required by an IOC need to be customized for each specific oil field development. Each technology and service embodies new concepts and recent technical progress. R&D in the supply chain is vital because the progress of technology and equipment is critical to the ability to locate deposits and increase the amount of oil that can be recovered. This customized production of technology and services is not present in the case of the automobile and aerospace industries, where the main company repeatedly produces the same product. In these industries, the system integrator negotiates with the supplier for the entire production and shares the production margin accordingly. On the supplier side, the knowledge that it will be supplying a particular piece of equipment to the core company for the whole production helps the supplier plan in advance and configure its manufacturing systems accordingly. The core company actively negotiates with the suppliers regarding the production process, timelines and specifications of the deliverables, and manages the supply chain in order to avoid any disruption during mass production. The main company must also make sure that suppliers are financially solid and will not file for bankruptcy in the middle of a production cycle.

Fourth, in the oil industry, the system integrator deals with much higher levels of uncertainty compared to other industries. In manufacturing industries, the system integrator is responsible for planning and assembling production. The risk that the system integrator takes is small, and uncertainty relating to the end product diminishes with each production step, starting from the conceptual and planning phase. In the oil industry, the system integrator plans, develops the field and bears the uncertainty of the subsurface. There is always a degree of uncertainty about the profitability of any given oil field. The system integrator, in our case the IOC, goes into production following exploration, appraisal and development phases. The process is long and highly capital intensive. After financing the exploration and appraisal stages and interpreting the subsurface data, the IOC must decide whether or not to proceed with drilling and developing the oil field, despite the uncertainty surrounding the subsurface and irrespective of how much capital they have already invested in the field. The 'final investment decision' (FID) is taken before the development of the field. Because the subsurface is largely unknown, the uncertainty is very high even at the FID stage. This level of uncertainty regarding the end product does not exist in many other industry sectors. Therefore, the role of the system integrator in the oil industry involves not only integrating the value chain but also taking the risk of the unknown.

Finally, in the automobile and aerospace industries, there are one or two main integrating companies. For example, in the aerospace industry, most of the suppliers work for Boeing or Airbus. In the oil industry, oil services can be provided to numerous companies including IOCs, but also independents or NOCs.

Despite all these differences in comparison to other industry supply chains, this book argues that the oil industry remains a good example of management by the system integrator.

2.2 TRANSFORMATION IN THE PETROLEUM VALUE CHAIN

Several structural changes have occurred in the petroleum supply chain over time. A few of these are outlined below.

Increased Outsourcing

Outsourcing in the oil industry began after the Second World War and increased over time following the oil shocks and progress in information technology (IT). During the '90s, IT capability became very important in drilling, reservoir management, modelling of complex systems and management of oil and gas flows in the system. Advancements in technology and IT capability made outsourcing increasingly more attractive to IOCs. Not only did enhanced IT systems ensure that outsourcing was easier and more efficient than executing business activity in-house, but it also delivered significant cost savings for the IOCs. Decreases in the market price of oil, compounded by sharp increases in the cost of extraction in the '90s, reduced financial returns that IOCs had previously realized by executing projects in-house. Outsourcing was seen as a solution to reduce the capital employed, during a period when cost reduction became essential.

Outsourcing has also been justified by the variable pattern of activity in the oil industry. For example, IOCs do not drill on a continuous basis. Drilling staff and systems are only required during the active periods of drilling. Outsourcing allows IOCs to use resources when needed and reduce costs through mutualization of the capacity. Accordingly, capacity, made up of assets such as drilling rigs and ships, is used by different companies according to their requirements. At the same time, costs are reduced for IOCs through competition among service providers.

Over the years, outsourcing has expanded for many activities in the oil industry, from managing systems to the construction of platforms. This has shifted a good proportion of IOCs' functions to OSCs. As the IOC business model evolved to an outsourcing-based model, highly qualified OSCs have developed thanks to the opportunities created by outsourcing. As a result, the oil services sector today is substantially larger than it was thirty years ago. Nowadays, only essential activities are kept in-house within IOCs and the majority of activities are outsourced to OSCs.

While the general tendency has been towards increased use of outsourcing (in areas like project management, construction of onshore and offshore facilities, engineering, etc.), there have been periods when IOCs realized they were outsourcing to the extent that they were losing essential competencies. Certain activities that were outsourced were later deemed unsuitable for outsourcing. These competencies were reintegrated back into the IOCs' central business model. One example, given by an industry expert, is that fifteen years ago IOCs were outsourcing the 'operation of the field' and 'maintenance of the field' during production. OSCs such as Schlumberger and Weatherford focused on gaining expertise in field development and operations and were eventually able to compete directly with IOCs. However, after realizing there was a tight market for field operations and recognizing that they were losing an important in-house capability, IOCs across the industry stopped outsourcing these activities.

In addition to the choice between outsourcing or in-house production, the degree of outsourcing for certain activities is also being questioned by IOCs. According to industry experts, while IOCs have increased the set of activities they perform, they have reduced their investment in engineering and in the management of construction projects. Owing to the growth of turnkey engineering, procurement and construction (EPC) contracts, they have cut staff numbers in the engineering and construction segment. However, according to an industry expert, several IOCs are beginning to see the limits of EPC contracts and conclude that they went too far in outsourcing in this area. EPC contracts are seen as suitable for well-defined and well-known technologies but are unsuitable for difficult fields requiring new technologies. A few IOCs agree that a certain level of engineering such as front-end engineering design should be kept in-house.

On the other hand, for some activities, such as drilling, while the outsourcing itself is not being questioned, the ability to control the outsourced activity and the responsibility in case of an accident are being reevaluated. In the past, IOCs used to own their own drilling rigs. They started to outsource the drilling activity in the '80s. There are many business drivers for outsourcing drilling that make it attractive to IOCs. For example, rigs used for drilling are an asset that needs to be fully utilized. If an international oil company possesses several drilling rigs per year, it has to operate them constantly. The IOC would have to manage its own rigs in all of its projects across the world, an activity which requires extensive and complex management. In addition, an IOC may not possess 100% of the rights for drilling according to its oil permit, because a majority of oil projects are done in joint venture (JV) with NOCs and other partners. Therefore, an IOC would have to justify the utilization of its own rigs to JV partners. IOCs might not have a sufficient amount of work to keep the rigs in use at all times. As a result, IOCs have chosen to outsource the drilling activity completely to OSCs. While in principle outsourcing drilling activities is an attractive option, recent events have led IOCs to question the control and the responsibility related to outsourcing. The BP oil spill during the summer of 2010 showed the risks of using rigs belonging to OSCs. A number of IOCs have since indicated that if the international oil company takes all the liability as the operator in case of an accident, then it should either execute activities to minimize the risks involved, or control all activities to a greater degree.

The activities that are outsourced vary according to each IOC. As an example, an industry expert stated that BP outsources basic engineering but Exxon keeps its basic engineering studies in-house. However, certain activities such as construction, drilling and well technologies are outsourced by nearly all IOCs. International oil companies agree that they do not need to have the in-house skills required to build a boat, a plant or a refinery. This would not only distract management focus, it would also be more costly if performed in-house. These activities are executed by OSCs according to the specifications given by IOCs. For example, IOCs design the Floating Production and Storage Units (FPSOs) at a high level but leave the construction of the FPSOs to the OSCs.

Continuous soul searching has taken place regarding the types of activity to be outsourced and the degree of vertical disintegration for the overall business, as well as for each individual activity. In the case of a new technology or complex geological context, the IOC prefers to directly manage, execute and control the business activity. IOCs are in consensus regarding keeping certain activities in-house, such as those which are 'close to oil' where the focus is on quality rather than cost. To wit, activities related to understanding of the subsurface and knowledge of the reservoir (such as geophysical interpretation and well design) are kept in-house because these inform the decision for further capital investment into an oil field. Once an IOC decides to invest and develop the field, the focus is on cost; hence it outsources related activities to reduce the costs. As an industry expert working for an IOC summarized, 'Haute couture is made internally; the standard services and products are outsourced'.

Outsourcing of an activity also depends on the geography and on human resources. Even though the IOCs may prefer to manage certain functions in-house, in some regions there is a shortage of qualified employees available. In such cases, IOCs must outsource activities based on availability of resources as opposed to cost or efficiency savings.

Regarding the integration of outsourced activities, companies like Total and Exxon are known to do the integration themselves. Other IOCs and a few independents prefer integrated projects. Because the costs and complexity of the fields have gone up, they prefer a unified system managed by an OSC.

All international oil companies agree that the ability to assess the work done during outsourcing is vital and should never be lost. According to industry experts, problems appear when a company loses the expertise to be able to check the work of its suppliers. IOCs need to retain sufficient competence and expertise to be able to evaluate OSCs and the quality of their work. Quality assurance and quality control of supplier services is essential, and the IOCs rarely outsource this activity. Furthermore, IOCs need to be expert on available and developing technologies in order to be able to choose the most appropriate technology during outsourcing.

Vertical Disintegration over Time

As stated previously, integration and disintegration of activities within international oil companies has changed over time. The IOCs were more vertically integrated before the 1950s, using their own rigs, tankers and technologies. For example, during the '60s Shell was internally operating everything from the canteen to the helicopters in Nigeria.

The increase in outsourcing and the development of capable oil services companies speeded up the vertical disintegration process for IOCs. Each IOC had to decide on the degree of vertical disintegration appropriate to its own company structure and adapt over time. In the 1990s, certain IOCs such as BP and Shell started disengaging from execution activities and focusing on asset management. The core of the oil business was seen as 'being the experts of the reservoir', more than 'being the master of the execution of the field development'. They began to be compared to portfolio managers, focusing on buying and managing oil fields, similar to the portfolio management function in a bank. However, after realizing that they were losing essential technical expertise and technological resources, they have re-internalized certain activities. For example, according to an industry expert, during 1990–2000 BP decided to outsource all activities except asset management in order to focus on exploration and long-term planning. Consequently, BP shut down engineering divisions and laboratories, replacing engineering teams with project managers. BP decided to be an 'informed buyer' not an 'engineer'. Over time, BP has become the most vertically disintegrated IOC in the industry. It has even outsourced activities that are seen as key expertise by other international oil companies, such as the interpretation of seismic data. According to an interviewee, BP made the seismic data available on the Internet and asked for interpretation by all service providers. BP informed companies that it would work with the service company that provided the best interpretation. However, following a series of high-profile accidents, BP has decided to reintegrate key engineering capabilities back in-house. For example, it chose to rebuild its refinery expertise in the US and set up laboratories (e.g., BP Institute in Cambridge) with the aim of regaining the ability to monitor and control contractors. Another example provided by the interviewees is that of Shell, which reduced its numbers of geologists and geophysicists in the late '80s but decided to re-internalize technological resources in E&P (exploration & production) and in geology in the first years of this century.

NOTES

1. Lease duration varies and it is usually awarded under two fiscal regimes: Production Sharing Agreements and Tax & Royalty Concessions.
2. Brent Field has produced nearly 10% of all the oil and gas extracted from British waters and gave its name to the most important globally traded blend of crude.
3. Physical separation is achieved through crude distillation in tall distillation towers, heating the oil to separate it into different groups of hydrocarbon compounds (fractions). Following distillation, the molecular structure of the separated fractions may undergo conversion in reformers, catalytic cracking units, alkylation units and other refinery equipment. Here heat, pressure and chemical catalysts break heavier oil elements into lighter ones (such as gasoline) or combine several light molecules into a few heavy ones for fuels such as high-octane aviation gas. Finally, many petroleum products undergo purification processes of some kind in order to remove chemical impurities. Blending involves mixing and combining of hydrocarbon fractions, additives and other components to produce finished products with specific performance properties (Hermann et al. 2010).
4. Petroleum pipelines, mostly underground, distribute crude oil and products to key regions. Tank trucks may be used to transport products to markets that are relatively close to the refinery. In addition to trucks, railroad tank cars are used to transport small-volume specialty products that are not economical or practical to ship by pipeline.
5. 'Petrochemicals are non-fuel compounds derived from crude oil and natural gas which take advantage of the reactivity of the carbon molecule and its

ability to create divers range of polymers which have different properties. All organic chemistry is based upon hydrocarbons (carbon based molecules) and derivatives of oil or natural gas and organic chemicals account for approximately 85% of all substances produced in the chemical industry—from basic plastics to complex pharmaceuticals' (Hermann et al. 2010, 189).

6. PFC Energy states that 77% of the world's 1.148 trillion barrels of oil reserves is controlled by governments (Blum 2005; Jaffe and Soligo 2007). Similarly, Jessen affirms that in the 1970s, international oil companies held approximately 85% of the world's known hydrocarbon reserves. However, three decades later, the IOCs' share of reserves has plummeted to less than 10%, with industry estimates ranging from 6% to 8%. Today, NOCs control as much as 94% of the world's oil and gas reserves (Jessen 2009). Howat (2006) also declares that the share of the integrated oil companies, known as 'super majors', has declined to lower than 10%; IOCs had access to around 5% of oil reserves and produced 16% of the world production in 2004 based on 80 Mb/day. According to Thurber (2012), oil majors and other IOCs have access to 9% and 18% of oil reserves but are responsible for 16% and 23% of oil production, respectively. Thus, NOCs controlled 73% of reserves and produced 61% of oil in 2009.
7. *Petroleum Intelligence Weekly* (PIW) compares petroleum-industry majors, independents and national oil companies based on operational size, rather than market cap or other financial measures, to provide a holistic view of the industry landscape. For the assessment, PIW incorporates six unique operational criteria: oil and gas reserves, oil and gas production, product sales and refinery distillation capacity. The 2011 list is provided by Broxson (2012, 9).
8. Standard Oil Co. of New Jersey v. United States Case, 221 US.1.
9. Standard Oil of New Jersey (Exxon), Standard Oil of New York (Mobil), Standard Oil of California (Chevron), Standard Oil Kentucky (Chevron), Standard Oil of Ohio (Sohio, which was purchased by BP), Standard Oil of Indiana (Amoco, which merged with BP), Continental Oil (Conoco) and Atlantic Oil (ARCO, which was purchased by BP).
10. Standard Oil New York (Mobil), Vacuum Oil Co. (Mobil), Standard Oil New Jersey (Esso), Anglo American Oil Co. (Exxon).
11. Aramco was later nationalized by the Saudi government and continues its existence as Saudi Aramco.
12. The 'Seven Sisters' was a term coined in the 1950s by businessman Enrico Mattei, then head of the Italian state oil company ENI, to describe the Anglo-Saxon companies that controlled the Middle East's oil after the Second World War. It is used to describe the seven oil companies which formed the 'Consortium for Iran' cartel and dominated the global petroleum industry from the mid-1940s to the 1970s. These companies were Anglo-Persian Oil Company (now BP), Gulf Oil, Standard Oil of California (SoCal), Texaco (now Chevron), Royal Dutch Shell, Standard Oil of New Jersey (Esso) and Standard Oil Company of New York (Socony, now ExxonMobil). The new Seven Sisters are NOCs such as Saudi Aramco, Gasprom, Petrochina, NIOC, PDVSA, etc. (Hoyos 2007).
13. The first of the 'oil shocks' of the 1970s was the 1973 oil crisis when Arab oil producers imposed an embargo. The decision to boycott America and punish the West in response to support for Israel during the Yom Kippur war against Egypt led the price of crude to rise from $3 per barrel to $12 by 1974 (Macalister 2011). The 1979 oil crisis followed the Iranian revolution. The nationalization of the Iranian oil industry and decline in Iranian oil production caused panic. The price of oil rocketed over the year from $15 to $39 and the rush to secure supplies caused acute shortages across the world (Baxter 2009).

14. Halliburton losses were $998, $820 and $979 million during 2002, 2003 and 2004 respectively (Halliburton 2004).
15. Drilling and well support accounts for 30 to 50% of an oil field's capital expenditure (CAPEX) and generates operating margins of 30 to 50% for the US service companies involved. In contrast, European engineering and construction activities involve huge CAPEX expenditure, but typically generate margins of only 6% or less. This structural difference in exposure to field CAPEX and margins assists in explaining why the US sector is more profitable (Hermann et al. 2008).
16. 'Cascade effect' is described as the cascading of concentration through simultaneous mergers and acquisitions in different layers of the value chain. First-tier suppliers of goods and services to global companies merge and acquire and lead to consolidation in their sector in order to develop leading global positions. These in turn pass on pressure upon their supplier networks. A fast-developing process of concentration has been observed in numerous industries supplying goods and services to system integrators (Nolan et al. 2005).
17. Exxon Neftegas Ltd. (ENL) has completed drilling the world's deepest well in the Chayvo oil field on the Sakhalin shelf in the Russian Far East. The shaft of well Z-44 is 12,376 metres deep (Russia Today 2012).
18. The threshold separating shallow water and deepwater can range from 656 ft to 1,500 ft (200 m to 457 m) in water depth. US authorities classify deepwater as water depths greater than or equal to 1,000 ft (305 m) and ultra-deepwater as water depths greater than or equal to 5,000 ft (1,524 m) (MMS 2008).
19. HPHT is formally defined as a well having an undisturbed bottom hole temperature of greater than 300 °F (149 °C) and a pore pressure of at least 0.8 psi/ft (~15.3 lbm/gal) or requiring a BOP with a rating in excess of 10,000 psi (68.95 MPa) (Schlumberger 2012).

REFERENCES

Baxter, K. 2009. "Ten Events in Oil's History that Shook the World." *Arabian Oil & Gas* [Online], 7 July. Accessed 5 December 2012. http://www.arabianoilandgas.com/article-5817-10-events-in-oils-history-that-shook-the-world/6/.

Blum, J. 2005. "National Oil Firms Take Bigger Role." *The Washington Post*, 3 August.

BP. 2011. "Summary Report 2011." Accessed 3 October 2012. http://www.bp.com/assets/bp_internet/globalbp/globalbp_uk_english/set_branch/STAGING/common_assets/bpin2011/downloads/BP_Summary_Review_2011.pdf.

BP. 2012. "Our History." Accessed 15 June 2012. http://www.bp.com/extendedsectiongenericarticle.do?categoryId=9039337&contentId=7036819.

Broxson, B. 2012. "National Oil Companies: Where Are We Now?" FTI Consulting, Houston Energy Group. Accessed 23 September 2012. http://www.haynesboone.com/files/Event/02c47aef-a062-470f-9136-296948be290f/Presentation/EventAttachment/36c9cee9-ec2b-4f3a-bf48-023c26343e4e/Bob%20Broxson%20White%20Paper%20Hou%20Oil%20and%20Gas%20Conference%20(2.pdf).

Chevron. 2011. "Annual Report 2011." Accessed 5 September 2012. http://www.chevron.com/annualreport/2011/documents/pdf/Chevron2011AnnualReport.pdf.

Chevron. 2012. "Saudi Arabia: Record of Achievement." Accessed 1 May 2012. http://www.chevron.com/countries/saudiarabia/recordofachievement/.

CNN. 1999. "Exxon-Mobil Merger Done." *CNN Money* [Online], 30 November. Accessed 1 September 2011. http://money.cnn.com/1999/11/30/deals/exxonmobil/.

ConocoPhillips. 2011. "Growing Value: 2011 Summary Annual Report." Accessed 5 September 2012. http://www.conocophillips.com/EN/about/company_reports/annual_report/Documents/ConocoPhillips%202011%20Summary%20Annual%20Report.pdf.

Economist. 2012. "The Unsung Masters of the Oil Industry." *The Economist*, 21 July.

ExxonMobil. 2011. "Summary 2011 Annual Report." Accessed 5 September 2012. http://www.exxonmobil.com/Corporate/Files/news_pub_sar2011.pdf.

ExxonMobil. 2012. "About Us: Our History." Accessed 1 June 2012. http://www.exxonmobil.com/Corporate/history/about_who_history_alt.aspx.

Finding Petroleum. 2012. "Andrew Gould—Take Advantage of Service Industry Competition." *Finding Petroleum* [Online]. Accessed 7 August 2012. http://www.findingpetroleum.com/n/Andrew-Gould-take-advantage-of-service-industry-competition/07029379.aspx.

Forbes. 2011. "The World's Biggest Public Companies, 2011." *Global 2000*.

FT. 2011. "FT Global 500 December 2011." *The Financial Times*.

Funding Universe. 2012a. "The British Petroleum Company plc History." *Funding Universe* [Online]. Accessed 1 September 2012. http://www.fundinguniverse.com/company-histories/the-british-petroleum-company-plc-history/.

Funding Universe. 2012b. "Chevron Texaco Corporation History." *Funding Universe* [Online]. Accessed 1 May 2012. http://www.fundinguniverse.com/company-histories/chevrontexaco-corporation-history/.

Goldstein, S. 2010. "Acergy, Subsea 7 Rally on $5.4 Billion Merger." *The Wall Street Journal*, 21 June.

Halliburton. 2004. "Looking Beyond: 2004 Annual Report." Accessed 5 December 2012. http://ccbn.mobular.net/ccbn/7/1132/1191/.

Hermann, L., J. Copus, and J. Hubbard. 2008. *A Guide to Oil & Gas Industry*. Global Markets Research. London: Deutsche Bank.

Hermann, L., E. Dunphy, and J. Copus. 2010. *Oil & Gas for Beginners: A Guide to the Oil Industry*. Global Markets Research. London: Deutsche Bank.

Hoovers. 2009. "Oil and Gas Field Services: Industry Overview." http://www.hoovers.com/oil-and-gas-field-services-/—ID__217—/free-ind-fr-profile-basic.xhtml.

Howat, I. 2006. Presentation. Finance and Investment Seminar. Edinburg, University of Stirling.

Hoyos, C. 2007. "The New Seven Sisters: Oil and Gas Giants Dwarf Western Rivals." *The Financial Times*, 12 March.

Hussain, R., T. Assavapokee, and B. Khumawala. 2006. "Supply Chain Management in the Petroleum Industry: Challenges and Opportunities." *International Journal of Global Logistics & Supply Chain Management* 1: 90–97.

Jaffe, A.M., and R. Soligo. 2007. "The International Oil Companies." The James A. Baker III Institute for Public Policy, Rice University. Accessed 24 January 2009. http://www.bakerinstitute.org/programs/energy-forum/publications/energy-studies/docs/NOCs/Papers/NOC_IOCs_Jaffe-Soligo.pdf.

James, R.A. 2011. "Strategic Alliances between National and International Oil Companies." Program on Energy and Sustainable Development, Stanford University: 104. Accessed 12 January 2012. http://iis-db.stanford.edu/pubs/23377/WP_104%2C_James%2C_NOC-IOC_Stategic_Alliances%2C_25_October_2011.pdf.

Jessen, R. 2009. "IOC Challenge: Providing Value Beyond Production." *Oil & Gas Journal* [Online], 107. Accessed 2 February. http://www.ogj.com/articles/print/volume-107/issue-5/general-interest/special-report-ioc-challenge-providing-value-beyond-production.html.

Macalister, T. 2011. "Background: What Caused the 1970s Oil Price Shock?" *The Guardian*, 3 March.

MMS (Minerals Management Service). 2008. "Deepwater Gulf of Mexico 2008: America's Offshore Energy Future."New Orleans: US Department of the Interior.

Nolan, P., D. Sutherland, and J. Zhang. 2002. "The Challenge of the Global Business Revolution." *Contributions to Political Economy* 21: 91–110.

Nolan, P., J. Zhang, and C. Liu. 2005. "The Global Business Revolution, the Cascade Effect and the Challenge for Catch-up at the Firm Level in China." Globalisation and International Business, MBA Module.

NY Times. 2015. "An Extensive Subset for the Brent Oil Field." *The New York Times* [Online], 18 February. Accessed 3 March 2015. http://www.nytimes.com/2015/02/19/business/international/royal-dutch-shell-dismantling-brent-oil-field-in-north-sea.html?_r=0.

Ochssee, T.B., C. Linde, J. Meijknecht, and T. Smeenk. 2010. "Competition and Cooperation of Economic Agents in Natural Resource Markets: A Dynamic Market Theory Perspective." *Polinares Consortium* [Online], 11. Accessed 23 November 2011. http://www.polinares.eu/docs/d1–1/polinares_wp1_dynamic_market_theory.pdf.

Russia Today. 2012. "Exxon Sets World Record with the Deepest Oil Well on the Russian Shelf." *Russia Today* [Online], 28 August. Accessed 5 September 2012. http://rt.com/business/news/exxon-sakhalin-well-record-727/.

Saipem. 2012. Presentation to Financial Community Preliminary 2011 Consolidated Results. *Saipem* [Online], 13 February. Accessed 17 March 2012. http://www.saipem.com/site/Home/InvestorRelations/artCatPresentations.2018.1.1000.4.1.html.

Schlumberger. 2012. "High-Pressure, High-Temperature." *Oilfield Glossary* [Online]. Accessed 16 May 2012. http://www.glossary.oilfield.slb.com/Display.cfm?Term=high-pressure%2C%20high-temperature.

Shell. 2010. "Shell Starts Production at Perdido." *Shell* [Online], 31 March. Accessed 12 June 2011. http://www.shell.com/global/aboutshell/media/news-and-media-releases/2010/perdido-31032010.html.

Shell. 2011. "Annual Report 2011." Royal Dutch Shell Plc. Accessed 26 September 2012. http://reports.shell.com/annual-report/2011/servicepages/welcome.php.

Shell. 2012. "Our History: The Beginnings." Accessed 1 December 2012. http://www.shell.com/global/aboutshell/who-we-are/our-history/the-beginnings.html.

Thurber, M. 2012. "NOCs and the Global Oil Market." Program on Energy and Sustainable Development, Stanford University [Online]. Accessed 15 March 2012. http://energyseminar.stanford.edu/sites/all/files/eventpdf/Thurber%20energy%20seminar%20NOCs%2006Feb2012%20final_0.pdf.

Thurber, M., and P. Nolan. 2010. "On the State's Choice of Oil Company: Risk Management and the Frontier of the Petroleum Industry." Program on Energy and Sustainable Development, Stanford University: 99. Accessed 15 May 2012. http://iis-db.stanford.edu/pubs/23057/WP_99,_Nolan_Thurber,_Risk_and_the_Oil_Industry,_10_December_2010.pdf.

Total. 2011. "Registration Document 2011." Accessed 5 September 2012. http://www.total.com/MEDIAS/MEDIAS_INFOS/5254/FR/TOTAL_Registration_Document_2011.pdf.

Total. 2012a. "The Challenges of Subsalt Imaging." *Total S.A.* [Online]. Accessed 3 December 2012. http://www.total.com/en/our-energies/oil/exploration-and-production/our-skills-and-expertise/the-deep-offshore/expertise/geophysics-201888.html.

Total. 2012b. "An Illustrated History of Total." Accessed 1 December 2012. http://histoire.total.com/index_en.html.

Total. 2012c. "Total at a Glance: An International Energy Provider." Accessed 5 September 2012. http://www.total.com/en/about-total/group-presentation/overview-total-940507.html.

UK Parliament. 2011. "Challenges of Deepwater Drilling." 6 January. Accessed 20 March 2011. http://www.publications.parliament.uk/pa/cm201011/cmselect/cmenergy/450/45005.htm#n1.

Victor, N.M. 2007. "On Measuring the Performance of National Oil Companies (NOCs)." Program on Energy and Sustainable Development (PESD), Stanford University: 34. Accessed 25 January 2009. http://iis-db.stanford.edu/pubs/21984/WP64%2C_Nadja_Victor%2C_NOC_Statistics_20070926.pdf.

Yergin, D. 1991. *The Prize: The Epic Quest for Oil, Money and Power*. New York: Simon & Schuster.

Yergin, D., and J. Stanislaw. 2002. *The Commanding Heights: The Battle for the World Economy*. New York: Touchstone.

Zhang, J. 2004. *Catch-Up and Competitiveness in China: The Case of Large Firms in the Oil Industry*. London: Routledge Curzon.

3 Impact of the Changes on the IOC-OSC Relationship

This chapter examines the relationship between IOCs and OSCs, summarizing key findings and concepts derived from in-depth interviews conducted with industry experts from IOCs, OSCs and NOCs. Because the contributions have been made in confidence by the interviewees, the chapter contains a limited number of specific references. It should, however, be considered as the overall outcome of all interviews.

3.1 RELATIONSHIP BETWEEN INTERNATIONAL OIL COMPANIES AND OIL SERVICES COMPANIES

3.1.1 Dynamics of the Industry: NOCs vs. IOCs vs. OSCs

As stated previously, national oil companies are resource holders and work in general as a JV partner with IOCs. The role of the NOCs changes according to their capabilities and the characteristics of the oil field. Certain NOCs such as Petrobras and Saudi Aramco possess similar capabilities to IOCs and prefer to work directly with OSCs in most projects. However, other NOCs such as Nigerian Petroleum and Iranian National Oil Company require the specific expertise of IOCs and form partnerships with them.

The nature of the relationship between NOCs and IOCs continues to evolve based on market dynamics. Traditionally, NOCs required IOCs to provide capital and take on the investment risks, for which they were rewarded with a share of the hydrocarbon produced. However, recently the increase in oil prices has meant that many NOCs have become capital-rich and started hiring OSCs with technological capabilities directly, bypassing IOCs. These developments have the potential to fundamentally change the business model of IOCs over time and reduce their lead role in oil fields.

As they function today, IOCs are similar to architects. The owner of a property decides to appoint an architect in order to construct a house. The architect then designs the house, specifies the technical details and tenders the work to appropriate service providers such as bricklayers, plumbers and electricians. Similarly, IOCs design the well, and contract drillers and well services companies. Hence in certain IOCs, there are business divisions called 'petroleum architects'. In the case of an oil field, the government provides acreage to the IOC, which then decides on methods of exploration, where to search for oil deposits, project timeline and methods of production, along with the services that are required for the successful outcome of the enterprise. Decisions on nearly all aspects of the exploration and production site, including decisions regarding drilling locations, health and safety standards and environmental protection measures, are the responsibility of the international oil company. In other words, IOCs are operators, financiers and project managers of the oil field. Due to the significant upfront costs, the cost of failure and the reward for success belong to the IOC.

Thanks to their understanding of risks throughout project life-cycle and their expertise within the field of reservoir knowledge, IOCs are experts in measuring, taking on and managing large-risk ventures. IOCs also do the political positioning and take on the political risk with each investment. The added value of IOCs is to make technological choices, integrate these choices, and finance, supervise and operate the project. In summary, IOCs have the proven track-record to finance, to assess risks and to manage large projects whilst working in challenging and politically complex countries.

The key to their success is their ability to search out the most highly skilled company for each service and integrate all services and technologies successfully. For example, an IOC may choose a different company for cementing (Halliburton), drilling (Transocean), logging (Schlumberger) and reservoir simulation (CGGVeritas), combining the best from each service segment into the project as a whole.

Another main characteristic of international oil companies is their presence within the whole value chain and their knowledge of the world market. For example, IOCs develop offshore fields and build liquefaction terminals for liquefied natural gas (LNG) in Qatar, while also maintaining a presence in gasification terminals in the UK. IOCs connect large resource holders to world markets. They deliver the end product—be it oil products, gas, LNG or LPG (liquefied petroleum gas)—to the markets.

While IOCs have transferred certain expertise to OSCs, certain capabilities that are judged essential such as exploration, reservoir evaluation, drilling strategy and oil architecture are kept in-house. When detailed, the in-house activities of IOCs form a long list of capabilities:

Geo-science: IOCs process and interpret seismic data, integrate all inputs and build up petro-models.

Reservoir: Knowledge relating to the presence and quantity of oil is the expertise of IOCs.

Drilling design: Methods of drilling and the design of drilling wells and their structures are decided by the IOCs.

Oil architecture: IOCs work as the architect of the oil field by developing its design and handling its overall project management.

Oil processing: IOCs deal with separation, compression, water injection and desulphurization. IOCs treat oil at the installation and separate it from sand, gas and water.

Technologies: Depending on the IOC, certain core technologies are researched, developed and kept in-house.

In general, long-term strategic planning and related decisions (such as choosing which fields to invest in, and portfolio structuring) are done in-house by international oil companies, while day-to-day operational activities are outsourced to OSCs.

OSCs, on the other hand, provide products, support and services to IOCs, NOCs, the independents, refineries and petrochemical companies. They execute a particular task on an oil field. OSCs are experts in specific technology, such as laying pipelines, drilling and seismic data gathering. Case studies in chapter 4 provide examples of specific tasks carried out by OSCs.

3.1.2 Business Models of IOCs & OSCs

The nature of the relationship between IOCs, OSCs and NOCs creates an interesting dilemma within the oil industry. OSCs supply services to international oil companies. At the same time, OSCs provide these services directly to NOCs. Consequently, there are some business areas where IOCs and OSCs compete with each other.

OSCs are typically seen as 'contractors'. While some interviewees, especially NOCs, see IOCs as 'super contractors', the business models of IOCs (operators) and OSCs (contractors) and the risks they undertake are completely different. IOCs invest capital, take an equity share in the oil field and search for return on equity (ROE), whereas OSCs use their equipment and technology and concentrate on EBITDA—Earnings Before Interest, Taxes, Depreciation and Amortization. In a nutshell, OSCs are technology providers assuming technical risks, whereas IOCs seek to master subsurface geology and take huge capital risks, in terms of both the reservoir and the price of hydrocarbons. Furthermore, because IOCs define their technology requirements and give specifications to OSCs, they are responsible for supervision.

Business structures of IOCs and OSCs and their interaction can be analysed from four different perspectives: revenue generation structure, business expertise and risk taking, areas of R&D and the use of technology. Each of these will be examined in turn below.

Revenue Generation Structure

The revenue of OSCs is generated by the services provided to customers such as IOCs or NOCs. Oil services company revenues are essentially the capital expenditure (CAPEX) of international oil companies (Hermann et al. 2008).

Oil services companies do not generate revenue in a way that is tied directly to the amount of oil and gas produced; their revenue is dependent on time spent and on the cost of the service provided. Most OSCs contract their work either by day-rate or on a 'meterage' basis (e.g., a fixed rate per day or a fixed rate per meter drilled).

Drilling services provide a good example of the revenue relationship between OSCs and IOCs. International oil companies do not own their own drilling equipment or employ drilling staff. Instead, they contract drilling companies to drill wells on a daily rate basis. Day rates for new rig contracts are announced by drilling companies. Contract drilling companies, therefore, generate revenue based on the amount of time they are contracted to work for IOCs (Harman 2007). As of December 2013 there were over 500 offshore working rigs worldwide, typically working on an average contract length of less than a year. There are contract announcements each month that provide valuable leading indicators on where oil industry costs and OSC revenues are heading. In the last years, the high price of oil has led to a surge in demand for all classes of rigs and has driven up daily contractor rates to record levels. As drilling demand increases, the cost of all other associated services (such as supply boats, helicopters, cementing, mud and wireline logging) also increases. IOCs often have many rigs working for them simultaneously, all at different rates and with different contract ending dates. Large increases or decreases in day rates take time to fully impact the cost base because there is a time lag between changes in day rates and their implementation across the portfolio of contracts (Hermann et al. 2008). Revenues of OSCs are therefore tied to the level of activity within the oil and gas industry, sometimes measured by the 'rig count' or the number of rigs working across the globe at any given point in time (Harman 2007).

Oil services companies generate revenue during the realization of the project. They need to generate revenue on each project during activity such as construction, drilling or seismic services. There is no upside potential once the activity is finished. Thus, the bulk of risk associated with oil exploration and discovery is borne by the IOCs, as oil service company revenues are only exposed to fluctuations in daily drilling rates.

Conversely, in the case of IOCs, the amount of oil discovered directly impacts their revenue. IOCs start generating income once the production of oil and its sale take place, and continue to do so over the years according to the production rate. The upside potential for their earnings is linked to the amount of oil as well as the price of oil. Because new reserves are the primary source of future revenue, IOCs invest significant effort and time into searching for new petroleum reserves. If an IOC stops exploring, over time it will naturally generate declining revenues from a finite number of depleting petroleum sources.

While OSCs do not generate revenue directly linked to the price of oil, they benefit from high oil prices just as IOCs do. For IOCs, the benefit is obvious; high oil prices mean the value of their assets and their revenue stream (which is directly linked to the price of oil) are worth more. For oil services companies, high oil prices also lead to increased revenues as large producers typically increase investment in new facilities in a high oil price environment. Some large oil companies increased their capital expenditure by 25 to 30% in 2007 and 2008 to bolster their chances of finding and exploiting new reserves. In a high oil price environment, rates as high as 1m$/day have been contracted to offshore drilling companies (Hermann et al. 2008).

The cost of oil projects is to a great extent the revenue of OSCs. International oil companies have a spending budget for each project, and they aim to reduce their capital expenditure (CAPEX) by qualifying several OSCs during the tender process and ensuring competition.

Business Expertise and Risk Taking

A high degree of uncertainty and risk is inherently present within the oil industry. The exploration phase requires relatively low capital expenditure; however it carries high risks due to geological uncertainty. Field development risk is also high due to greater capital requirements despite reduced geological uncertainty (Thurber and Nolan 2010). Governments require IOCs to bring capital and technology, bear the risk of finding oil (exploration), develop the oil field and manage the production process. International oil companies take the risk of reservoirs yielding little or no oil, losing invested capital and associated risks arising from combining the required technologies.

Risk and Expertise of the Reservoir: As stated previously, petroleum crude oil is found in tiny pores of sedimentary rocks which typically lie both underground and under the sea, sometimes as deep as 3 kilometres below surface. Even with today's technology and expertise, only one in five wells leads to discovery of commercially viable reserves. Furthermore, even when oil is found, its quantity within the reservoir is subject to interpretation. The risk taken during the search for oil, the expertise of accessing the acreage and assessing it correctly are the main expertise of IOCs.

IOCs have geological models to determine and evaluate the presence of oil. They benefit from possessing in-depth expertise as a result of working with diverse geological data. Results from previous exploration and production activities enable them to compare the characteristics of different geologies, leading to better analysis of new fields. Hence, IOCs benefit from an economy of scale regarding knowledge. As described earlier and as will be further detailed in the seismic services case study, reservoir knowledge is a core competency of IOCs and is kept in-house, even when the seismic acquisition is outsourced to seismic companies.

Once oil is located, the IOC assesses the amount of oil that may be present under the ground and defines prospects and exploitation conditions accordingly. In other words, IOCs choose where to explore, assess the specifications of the reservoir and decide on the drilling strategy for the reservoir.

Because reservoirs cannot be seen with the naked eye, they are always subject to interpretation. Therefore, being able to assess the state of the reservoir and determine the quantity of oil it is expected to produce is a key expertise of IOCs. Competence and knowledge in the field of geo-science enables the IOC to assess the profitability of a reservoir, which in return drives the profitability of the IOC itself.

On the other hand, OSCs are indifferent to the amount of oil found in the reservoir as their income and profits do not depend on it. They are not experts in this field, because the risks linked to reservoir yield, surface and exploration do not impact them, and neither do they seek to exploit the oil. Consequently, they do not want to take the risk of the 'unknown'. OSCs provide IOCs with the tools such as mud logging and seismic data gathering, which enable the IOCs to interpret geological data. By using these tools, the IOCs are able to determine whether or not to invest in the reservoir.

Risk and Expertise of Investment: International oil companies risk high capital expenditure in order to explore oil fields. There is no guarantee that reservoirs or oil fields will yield sufficient oil to recuperate investment costs. The capital expenditure for new oil fields amounts to billions of dollars. For example, the total cost of the Girassol project in Angola has been $2.8 billion (Total 2003).

As new oil fields require a significant outlay of capital, IOCs must strategically identify which oil fields to invest in. Investment decisions cannot be taken in isolation based on the cost and benefit of a certain acreage; rather, IOCs make their investment decisions in relation to their entire portfolio of investments. At any one point in time, IOCs are simultaneously managing and reviewing investment opportunities in multiple locations around the globe. Their ability to effectively manage these activities in parallel is central to their profitability.

Oil services companies do not invest in the oil field, these risks being undertaken by IOCs. In the words of an interviewee, 'OSCs offer services for the projects and invest in the means but not in the final outcome'. Because OSCs work on a service-fee basis and generate revenue during the realization of the project, they do not carry similar levels of capital risk compared to IOCs.

Risk and Expertise of Combining Technologies: International oil companies work as 'maître d'ouvrage'. The majority of OSCs are specialized in individual techniques, while the IOCs' expertise is employed in integrating these individual activities into sophisticated models, because IOCs understand the pros and cons of key technologies. For example, many tools are available for exploring the oil field such as logging techniques, rock doctors and seismic methods. IOCs work to decide which methods and tools to use and then integrate those selected in order to determine whether further investment in exploring the oil field is viable. In effect, they combine all available relevant technology and manage the venture as one big project.

IOCs' expertise lies in finding the best combination and optimum structure for oil field development. For example, in drilling, rig companies contract the machine and the staff to an IOC but the decision relating to how the well will be drilled and what kind of equipment will be used belongs to the IOC in question. The knowledge lies in the specifications that IOCs provide and in decisions relating to the choice of the technology and development concept that will be employed. International oil companies are instructors; they provide direction and pace on an oil field and act as a system integrator. They function as the coordinator of the hub of activities on an oil field and ensure successful interaction of the different firms involved. OSCs, on the other hand, work according to the specifications given by IOCs.

The majority of oil services companies specialize in particular technologies and rarely offer integrated technologies. Even OSCs that possess a wide range of services and technologies do a limited amount of integration, with the exception of few such as Schlumberger, which offers an 'integrated project management' service.

Design of the Oil Field: IOCs decide on the concept of the oil field[1] and provide service companies with their specifications, investment decisions and technical requirements, which the OSC is then responsible for applying. For example, IOCs will specify the precise length, diameter and material required for the pipeline in the oil field. The IOC determines what specification is most suitable for the oil field in question, and the OSC builds the pipeline to specification.

Areas of R&D and R&D Expense

Among the current challenges for the oil industry are the technological limitations and the depletion of natural resources. As oil fields continue to be developed, new oil reserves can only be found in deeper, hotter and more technically challenging areas. There are unique challenges to every oil field, and therefore IOCs and OSCs spend a significant amount of time searching for solutions to problems particular to each project. Although IOCs and OSCs work together on the required technology, their areas of investigation and their R&D budgets highlight differences between them.

IOCs tend to focus their R&D efforts on investigating and assessing the feasibility of new concepts in oil exploration and production technology. Therefore, their research addresses general scientific problems and deals with more open-ended questions. Once a new concept is developed, IOCs tend to work together with OSCs, which then carry out further research to develop the necessary equipment and materials that translate the concept into practical application. For example, as mentioned by an interviewee, the R&D division of an IOC carried out extensive research to understand the relationship between the salinity of the water which is injected into the field and the amount of oil that can be recovered as a result. While this type of research may not have immediate practical application, it may lay down the foundation for future equipment and services.

In this regard, it may be considered that IOCs outsource many R&D tasks to OSCs with the exception of certain areas, such as subsurface, where IOCs keep R&D in-house and develop highly confidential technology.

Strict confidentiality and security measures are often implemented for these critical R&D functions. Employees may be restricted from key project knowledge, with 'Chinese walls' separating various members of the research team. In an example given by an interviewee, while developing a new chemical compound in an IOC, each individual worked on a separate component so that only a few individuals knew the complete formula of the new compound.

In contrast to IOCs, the R&D of OSCs concentrates on developing specific technologies with practical applications, tools and materials. They tend to do research on 'specific points' and not on broad concepts like IOCs. For example, an OSC may search for the best materials for HPHT environments.

While focusing on different R&D areas, IOCs and OSCs conduct complementary work and develop research in a coordinated manner. For example, the flexible riser concept developed by Total became an IPB (integrated production bundle)[2] in the hands of Technip. Following the Total concept, the IPB technology was first and successfully deployed by Technip on behalf of Total on the Dalia field, offshore Angola, at a water depth of between 1,200 and 1,500 metres. Today, flexible pipe and riser technologies are among the prime focus areas of R&D in Technip (Technip 2012). Another example is the Wide Azimuth[3] technology (CGGVeritas 2012). BP developed the concept but was unable to produce it due to equipment shortages and the absence of economies of scale. The technology was consequently developed by CGGVeritas at the request of BP based on BP's original idea.

Comparison of the amount of research and level of investment between IOCs and OSCs sheds further light on the nature of relationship between these players. Historically IOCs were larger investors in R&D in comparison to OSCs—which led, for instance, to the commercialization of North Sea oil fields. However, today, number of patents and other indicators show that OSCs are more prominent incubators of new technology. As seen in Table 3.1, Schlumberger, Halliburton and Baker Hughes filed more patents than most IOCs, based on the list of patents published by the Patent Board in the *Wall Street Journal*[4] (Patent Board 2012). A high level of patenting is observed in certain IOCs such as Exxon, which also has the highest Science Strength rating, indicating a close connection to core scientific research (IPIQ 2012).

Research and development is also being undertaken in collaboration with other institutions. Both IOCs and OSCs have collaborative R&D departments that coordinate with universities and institutions such as IFP and University of Cambridge. For example, IFP and Total have created a research group in sedimentology and reservoir characterization at IFP School. Most research will be carried out in IFP's laboratories or in Total's R&D centres. Schlumberger has founded the Gould research centre next to Cambridge University and benefits from strong collaborative links with the university to drive innovation (Schlumberger 2012c). The University of

Table 3.1 Top Ten Innovators in Energy and Environmental Patent Scorecard

Company	Current Rank	Previous Rank	Patents Granted	Science Strength™	Innovation Cycle Time™	Industry Impact™	Technology Strength™	Research Intensity™
Shell	1	2	148	1588.5	23.6	4.77	566.44	1.48
Schlumberger	2	1	581	1041.3	13.9	1.2	556.09	1.05
Halliburton	3	3	230	2745	12.7	1.98	362.6	2.25
Baker Hughes	4	4	355	835.3	13.8	1.28	361.4	1.59
General Electric	5	6	175	50.8	13.6	2.32	323.26	0.5
ExxonMobil	6	5	321	7778.3	14.1	1.13	288.41	2.87
Vestas Wind Systems	7	9	78	6	13.6	2.54	158.69	0.08
Weatherford International	8	7	105	252.3	16.7	1.65	138.95	1.29
Chevron	9	8	158	1352.8	14.3	0.94	117.82	2.44
Siemens	10	11	149	9	11.5	0.93	111.01	0.15

Source: IPIQ 2012; Patent Board 2012.

Note: **Quarterly snapshot, thirteen-week averages.*

Cambridge BP Institute is another example that supports research focusing on fundamental problems in multiphase flow and is highly interdisciplinary, spanning six university departments such as Earth Sciences and Physics (BPI 2012).

Finally, IOCs and OSCs differ in terms of R&D expenses. Comparing R&D figures may be misleading because the definition of R&D and the classification of expenses under R&D differ between companies. With this caveat, looking at the current R&D figures contained in annual reports, we can conclude that the R&D budgets of large OSCs compete with those of the IOCs (see Table 2.4). For example, Schlumberger's R&D expenditure was higher than all the IOCs except Shell,[5] which matched Schlumberger's spending of $1.1 billion in 2011 (Shell 2011c). The contrast is even more striking if comparison is done based on R&D expenses as a proportion of revenues. Overall, IOCs do not match the service sector's spend on R&D as a percentage of revenues, with IOCs spending less than 1% of revenues while the service companies spend 3 to 4% (Thuriaux-Alemán et al. 2010). For example, in 2007 Shell and Exxon spent 0.3% and 0.2% of their revenue on R&D respectively[6] (Crooks 2008).

Use of Technology

The oil industry is one of the most technologically advanced industries in the world. In addition, because most projects are done with JV partners, it is an industry where technology spreads very quickly. Intellectual property is hard to keep under wraps for more than two to three years due to the partnership nature of most projects. In most cases specific technological advances are made by OSCs with concepts provided by IOCs.

OSCs are highly technology-focused companies with employees specialized in various specific technologies (e.g., well servicing specialists), whereas IOCs employ individuals with broader capabilities and knowledge. When IOCs require a specific technology or a workforce with specialist knowledge and qualifications, they organize tenders to OSCs with expertise in that area. A key driver of IOCs' success in the industry lies in their ability to find and apply technological advances. BP has a Technology Advisory Board with a mandate to look for technologies being developed outside of BP. According to an interviewee, while IOCs may have limited internal R&D budgets, they have access to universities and other small technology companies with the aim of developing their ability to access changing technologies. BP's and Total's collaborations with University of Cambridge and IFP respectively are examples of IOCs accessing university-based research and technical knowledge. IOCs also form venture capital funds which enable them to invest in innovation by providing seed funding to companies developing new and innovative technologies. Their objective is to 'buy technologies' and benefit from the first-user advantage. Shell Technology Ventures Fund is an example of a fund that focuses on the development and deployment of new technologies.

Needless to say, the adaptation of new technology varies between IOCs. Some IOCs are quicker to adapt and implement new technologies, whereas others take a more conservative and cautious approach.

Finally, while IOCs invest less in R&D than do OSCs, as system integrators they guide the process of technical innovation as they operate in technically challenging areas with intense pressure on margins from profit-seeking private shareholders. Technological progress in the supply chain is therefore mainly driven by international oil companies.

3.1.3 Fundamentals of the Relationship between IOCs and OSCs

International oil companies operate as systems integrators and manage the pace and scope of their relationship with OSCs. The relationship is highly complex based on the significant number of interdependencies. OSCs provide critical services to IOCs. If oil services companies are unable to deliver on their contractual obligations, the IOC's business will suffer.

The interpretation of the relationship varies. In an interviewee's words, 'The relationship is similar to an unformulated partnership with permanent friction: continuously up and down'. Another interviewee from an OSC said, 'There are cycles in the relationship. IOCs sometimes find the market share of a particular OSC too high and try to rebalance it by reducing the amount of work contracted. When its market share decreases, IOCs reevaluate the situation and reestablish the business relationship with the company.'

According to most interviewees, a few factors such as consolidation in the services sector and oil price cycles significantly affect the relationship between IOCs and OSCs. First of all, consolidation in the market structure of OSCs has impacted their relationship with IOCs. Some IOCs believe that service sector consolidation has resulted in the most efficient and robust OSCs remaining in the industry. These OSCs have the ability to take on larger and more complex tasks and have become powerful counterparts for IOCs. In contrast, other IOCs believe that the concentration of technology, equipment and services in the hands of a few large OSCs has significantly restricted the choice of suppliers and caused IOCs to lose negotiation power with their subcontractors. A few interviewees mentioned that in some cases there is only one OSC available to provide specific equipment. For example, an interviewee estimated that Schlumberger alone provides 70% of the logging services. There are also cases where IOCs are 'obliged' to continue working with the same company for the follow-up of services. For example, in the case of BOPs, from the IOCs' perspective it is better if maintenance and spare parts are provided by the original supplier. In all these cases, there is little incentive for the OSCs to offer favourable terms to the IOCs. Following consolidation, some IOCs now believe they had become too dependent on a small number of contractors and went too far in transferring responsibility, expert knowledge and savoir-faire.

Second, the relationship between international oil companies and oil services companies along the supply chain has been very cyclical. Oil prices and the overall macro-economic situation have been the most important factors behind the cyclical relationship. High oil prices (e.g., the 2004–2008 period) intensify oil exploration and development activities, increase the work load of OSCs and reduce available capacity. When oil prices are high, OSCs can demand higher prices for their services and negotiate better terms and conditions in their contracts. In this environment, the risk–profit balance moves in favour of subcontractors. Conversely, when oil prices drop, IOCs are not in a hurry to evaluate and explore new oil fields. This reduces the demand for services and puts pressure on OSCs to reduce their prices. As IOCs place increasing pressure on OSCs, oil services companies in turn exert more pressure on their suppliers. This cascade effect often impacts the entire supply chain. Therefore, the pressure applied by the IOCs on the OSCs can have a negative impact on the profitability of smaller companies further along the supply chain.

In other words, volatile oil prices put a unique tension on the relationship between IOCs and OSCs. With each oil price cycle, both IOCs and OSCs seek to gain a competitive advantage. For example, the balance of power shifted in favour of OSCs following the run up in oil prices between 2004 and 2008 when the IHS CERA Upstream Capital Cost Index more than doubled from 109 to 230[7] (IHS 2015). The volatility in oil service prices reached a peak in 2008. In the first half of that year, OSCs used all of their resources and had little spare capacity available for ad hoc requests from IOCs, consequently charging high fees. In the latter part of the year, with the onset of the financial crisis and significant drop in oil prices, OSCs had to cut their prices due to decreased demand for their services. The balance of power once again shifted in favour of IOCs.

Although fluctuating prices can create tension in the relationship between IOCs and OSCs, in some cases changing oil prices may also lead to further collaboration due to the knock-on effect oil prices have on all companies concerned. A good example which was provided by an interviewee is where an IOC helps an OSC to reduce its costs so that it can lower its fees without significantly affecting its profitability. OSCs typically have lower credit ratings than IOCs and therefore pay higher interest to borrow money. The cost of borrowing is then reflected in the price of their services. The IOC in question proposed reducing the cost of borrowing for the OSC by paying in advance or improving the payment schedule. A similar example revolves around foreign exchange. Prior to 2008, an IOC requested all quotes for an OSC's services in US dollars. Therefore, in order to be able to offer all prices in USD, the OSC had to hedge against currency fluctuations by purchasing currency products from banks. However, as IOCs have multiple income streams in different currencies, their need to cover their currency exposure is lower and cheaper than that of OSCs. In order to help OSCs to reduce their costs, the IOC in question stopped asking for all services to be priced in USD.

However, despite these examples, a few OSCs have complained that following the economic crisis in 2008, the relationship with IOCs has become bitter and more difficult to navigate owing to the high cost-containment pressures.

Not all services are affected in the same way during an oil price decrease or economic crisis. For example, drilling and engineering contracts are long-term (five to ten years) capital-intensive contracts compared to, say, well services contracts. Hence they are less exposed to short- to mid-term fluctuations in oil price. As an example, the construction of an FPSO takes place under a five-year contract where the prices are fixed. Because IOCs could not change the price for these contracts, they focused on shorter term contracts such as those offered by well services companies following the economic crisis in 2008. Therefore, the burden of reducing costs during this period was borne primarily by well services companies.

The relationship and balance of power between IOCs and OSCs affects their respective profits, contractual risk, contractual responsibilities and liabilities. The relationship determines the level of risk and contractual exposure each company is willing to assume. For example, the balance of power during a contractual negotiation could determine who will undertake the weather risks (i.e., who pays for the waiting time that arises due to adverse weather conditions) or liabilities such as pollution. While it is ultimately the IOCs' responsibility to run the projects in an environmentally friendly manner, the specific contract between IOCs and OSCs could make OSCs also responsible for preventing pollution and sharing the costs arising from pollution caused by an accident.

As discussed above, high volatility in oil prices strains the relationship between IOCs and OSCs. Because both IOCs and OSCs would benefit from a stable balance of power, there is a growing trend within IOCs to seek strategic relationships with a relatively small number of OSCs through long-term contracts. A publicly available example of this trend can be seen in the recent transaction between Shell and Transocean. According to the deal, Transocean will provide four newly built ultra-deepwater drill ships to Shell for a period of ten years (Team 2012a).

Structures of Collaboration between IOCs and OSCs

Different structures of collaboration have taken place between IOCs and OSCs. The '90s was a period of alliances between the operator (IOCs) and the subcontractors (OSCs), with the aim of putting together the required skills needed for key projects at the lowest total cost. For example, the Andrew field in the UK North Sea was developed based on the alliance model. The Andrew alliance consisted of the operator BP Exploration, Schlumberger IPM for well management and data acquisition, Baker Hughes INTEQ for integrated drilling services, Transocean for mobile rigs and Santa Fe for platform rigs (Bourque et al. 1997).

Subcontractors were asked to form an alliance with commercial agreements laying out the distribution of risks and profits. For the subcontractors, there were big bonuses if they delivered the required services on time; however, there were also large financial penalties for delays. Payouts for contractors were very lucrative, but these could be easily lost. For example, a twenty-four-hour delay during horizontal drilling could result in a large penalty. The collaborative development model of alliances disappeared over time due to the mistrust between parties. Inherently, contractors and operators have conflicting interests; the contractor wishes to maximize profits while operators aim to minimize their expenditure. Negotiating and target setting for completion dates or required efficiencies proved to be difficult. The main shortcoming of the alliance model was that it failed to develop common objectives between contractors and operators. Furthermore, the cost of alliances was often higher than the market prices for the services involved. Due to its shortcomings and the rising strength of IOCs' procurement departments, the alliance model became obsolete in the early 2000s.

Current structures of collaboration vary according to commercial contracts between IOCs and OSCs. There are several contract types managing the relationship between IOCs and OSCs in the petroleum value chain. The main ones are fixed-price contracts, such as turnkey contracts and EPC contracts, and variable-price contracts, such daily rate contracts and cost-plus contracts.

In *turnkey contracts*, the contractor and the customer agree on a fixed lump sum amount for the entire project. Because the turnkey contracts work on a fixed-price basis under which OSCs assume weather, geographic and performance-related risks, they are not favoured by OSCs. From IOCs' perspective, turnkey projects are good for standard projects. They provide stronger cost control incentives and a more predictable bill. However, they do not provide sufficient incentives for innovation and hence are not suitable for the non-standard projects that require innovative thinking and new technologies. Where innovation is required, the cost-plus or partnership models are considered more suitable.

Turnkey contracts are only suitable for certain services. It is very difficult to contract exploration projects for a lump sum because the requirements of the activity and subsurface conditions cannot be fully assessed at the start of the project. For example, it is not known how many metres will need to be drilled to reach oil formation so it is therefore unwise for an OSC to accept a fixed-price contract. Alternatively, turnkey contracts are specifically useful for projects whose scope can be defined in detail in advance, such as the construction of a refinery.

Engineering, procurement and construction (EPC) contracts are a specific type of turnkey contract used in engineering and construction projects where the OSCs take the entire project risk. The engineering and construction service company coordinates all activities such as design, development, construction, transportation and installation of equipment and manages all related procurement.

In *cost-plus contracts*, OSCs charge the cost plus a pre-agreed margin to IOCs. Generally, these contracts exist in pure engineering services rather than engineering and construction services.

Under *daily rate contracts*, oil services companies charge a fixed daily rate to IOCs. These are often used in drilling where OSCs are remunerated by daily drilling rates based on the location, type of rig used and prevailing market conditions. Another type of contract that is similar to daily rates is one based on the depth of the well drilled. These are called 'meterage' or 'footage' contracts. These are less desirable for OSCs because the speed of drilling can be affected by many factors beyond the control of the OSC (such as weather conditions). Therefore, it is risky for the driller to accept a meterage contract. A difference exists between payment per metre drilled (unit rate) and per day (time rate); the former is similar to the fixed-price model and the latter to reimbursable contracts (Osmundsen et al. 2009).

The relationship between OSCs and IOCs varies based on the type of contractual structure. For example, the ability of IOCs to influence the activity of an EPC company is largely reduced if parties are working on a lump-sum turnkey contract. In practice, this will often mean that the IOC must cede influence during the operation. The fixed-price contracts are more likely to produce delays as they involve bureaucratic process each time a change is required. Where a fixed-price contract is concerned, it is difficult to make changes to the contract once it is awarded (Osmundsen et al. 2009).

Reimbursable contracts provide weaker cost incentives and a more uncertain final price. But it is easier for the operator to secure changes and influence the work process. This represents a trade-off from the IOCs' perspective (Osmundsen et al. 2009). IOCs tend to be more in control of project activity when the contract is completed on a cost-plus model. In some contracts, IOCs dictate the project requirements and all the specifications for the items to be purchased. The OSC receives quotes for prices of the goods and communicates these to the IOC, which then chooses the supplier.

In some cases, the project starts on a cost-plus basis until it is clearly defined and continues as fixed-price contract. The cost-plus model is used to finance initial feasibility studies and basic engineering studies. Once the work has become clearly defined, risk may be transferred to OSCs and the contract terms and conditions change to a 'fixed-price' or 'lump-sum' contract. In this case, it is imperative that OSCs work very closely with IOCs during the definition phase in order to understand the full scope of requirements and risks of the project. OSCs are then able to finalize the procurement and propose a lump sum for the remaining work.

International oil companies may choose to integrate all services themselves or request an integrated package. During the construction of an oil platform, IOCs may split the platform into smaller projects and manage the integration or opt to subcontract the entire build of the platform. For example for an FPSO, engineering design work, building work and subsea segments may be given to different companies. Alternatively, the whole construction of the FPSO can be handed over to one company. In each case, the relationship and responsibilities of IOC and OSC will vary. When IOCs opt to manage the integration of each component of the project, they apply a 'pick and mix' approach. The IOC goes to the marketplace for each of the services required and creates a short list of preferred bidders based on their technical capability, price and experience. On tenders, suppliers compete with one another and develop their technical and commercial offers for the services. The suppliers are often 'prequalified' according to their technical offers. Commercial offers are evaluated afterwards. Competitive tenders and reverse auctions take place between the short-listed companies, and arm's length transaction takes place for each service. Then, the IOC manages the final integration of all procured services.

The structure of collaborations between IOCs and OSCs also differs according to geographical regions due to prevailing business culture, availability of well-established infrastructure and reliable service companies. For instance, in Europe, IOCs prefer more integrated services and tend to give more responsibility to contractors. In the US and the Gulf of Mexico, IOCs operate as project managers and often hire different contractors for each specific service. In other areas, such as West Africa, the Far East and the Caspian Sea, infrastructure is more limited. IOCs prefer to have one main contractor and tend to use turnkey contracts.

Day-to-Day Work

The level of involvement of IOCs in the day-to-day work of OSCs varies according to the model of collaboration and the type of activity. International oil companies may have less involvement during the construction of the platform in a building yard but be much more involved in the day-to-day drilling activity on the oil field.

As operator, IOCs have ultimate liability for the oil field because they are the responsible decision-makers for the project. The liability of OSCs is set forth in the contract and is typically capped by a predetermined amount.

Interviewees from several OSCs confirmed that IOCs are in control of the majority of decisions in the oil field on a daily basis. IOCs specify the tasks and activities that will be completed and sign off on key decisions. All activities are reported to and confirmed with IOCs. An example is given in the well testing area: well site engineers of IOCs propose an approach for well testing, set targets and liaise with the OSC to confirm if these targets are achievable. Continuous communication and liaison between both parties ensure that the project is achievable and in line with expectations. The IOC continuously monitors the operations and the on-site progress of the OSC. In the well testing area, the OSCs carry out buildup, drawdown and multi-rate testing to provide IOCs with the necessary information such as temperature, pressure and flow rate to qualify the reserves as 'proven reserves'. Reservoir engineers and petro-physicists of the IOCs analyse the data and evaluate the potential value of the reserves. IOCs specify the data requirements, and OSCs ensure gathered data meet these requirements.

Another example of the control of day-to-day activities by IOCs can be seen in seismic services. The IOC sends a statute of requirements to the OSC in which it indicates all specifications such as dimensions, material to use and the surface of investigation for the seismic activity. During seismic data gathering, there are weekly or bi-weekly meetings between the IOC and the OSC to review progress. The seismic company collects the data and carries out necessary tests specified by the IOC. The IOC gives final approval for each stage of data gathering and testing.

As contract holders, IOCs control the activities of OSCs to ensure the quality of the OSCs' work and ascertain that progress is being made according to contractual timelines. An oil field is an environment of great uncertainty and risk, and therefore close cooperation in daily business is considered very important. Furthermore, the involvement of IOCs in daily work facilitates better understanding of the challenges that work presents. IOC teams work on the project alongside OSC teams both on- and off-site. There are also cases where OSCs have employees in IOCs' offices. For example Schlumberger has employees working in Total's office to assist with IT software.

For some projects, IOCs use an external company to review the work of the OSCs. External consultants may work on the OSCs' site on behalf of IOCs. However, several OSCs mentioned that they prefer to work directly with the IOC engineers rather than consultants as the presence of consultants may complicate the relationship between IOCs and OSCs by creating an additional layer.

Relationship within the Context of Joint R&D and Development of Technology

Joint R&D and development of technology is a prime example of the deep and close relationship between OSCs and IOCs.

IOCs use new technologies to gain competitive advantage and access to new oil reserves. There is a continual need for international oil companies to develop new and improved technologies in order to remain forerunners in the industry. Although they are significant users of new technological progress, IOCs have pushed practical R&D and the creation of new technologies onto oil services companies for several reasons.

The first reason is commercial. Due to their specialization, OSCs are better positioned to benefit more from a new technology within their area of expertise. Because OSCs can utilize the new technology in more projects in any given period, they have the potential to generate higher returns on their investments compared to IOCs. In addition, OSCs can further develop and refine the new technology quicker by leveraging feedback received from multiple clients.

Furthermore, the benefits of technological development in the oil sector are often not realized over a short period. Cost and efficiency savings often take many years to be fully realized. There are certainly a few breakthrough and game-changing technologies in the oil industry, but most technologies represent incremental progress. Moreover, new technology spreads very quickly thanks to JVs in each oil field. If an IOC develops a new technology, it is difficult to keep it for its own exclusive use, owing to the nature of the industry and the close working relationships that take place within it. Therefore any competitive advantage based on a new technology is likely to disappear quickly.

As a consequence, IOCs have pushed practical R&D and the creation of new technology onto oil services companies. Hence, OSCs have started initiating the development of technology, as well as assuming the risks involved. However, as discussed, while the progress of new technological development in the oil industry is now largely carried out by OSCs, IOCs remain engaged in the process and have considerable influence on the evolution of new technologies, guiding the R&D efforts of OSCs.

The influence of IOCs on OSCs regarding the development of new technology can be analysed from multiple aspects. First of all, new concepts in the oil industry such as deepwater exploration, high temperature or unconventional fields are often formed by IOCs. Each time an IOC explores an oil field that presents unique challenges, it researches and investigates the best approach for harnessing the natural resources of that oil field. For example, IOCs investigate the most appropriate technology to implement in the harsh operating conditions in the Arctic region. IOCs tend to explore in deeper, hotter and more technically challenging areas. These technical challenges in field development drive and determine the R&D priorities for OSCs. While IOCs do not typically dictate explicit strategic priorities for OSCs, they influence OSCs to prioritize certain technologies.

In certain cases, IOCs generate an approach for improving the quality and durability of equipment and processes, but do not possess the technology or economies of scale in-house to finance the research required to develop their concepts further. They then partner with OSCs under research agreements in order to develop these ideas. For example, Total, Chevron and Schlumberger jointly developed INTERSECT, a next-generation reservoir simulator (Schlumberger 2012b).

Sometimes the influence of IOCs may be more direct. For instance, they may directly request new technologies or services from OSCs by creating a request for a proposal (RFP) for a specific requirement. OSCs then formulate a solution to meet the new requirements and develop the necessary equipment, technology and processes. For example, Saipem created a new technology in its Blue Stream project as a direct response to demand from ENI and Gazprom. The operator had recognized a shortcoming with the quality and durability of pipes during a key phase of a project. Saipem developed and proposed the J-Lay solution, which reduced the stress placed on the pipe and increased reliability. Saipem's collaboration with the operator was essential in order to allow Saipem to understand the precise needs and requirements and develop an appropriate solution to meet them.

From the perspective of the oil services companies, the ability to remain competitive in the market relies on investigating and developing new technologies which meet their customers' current or future requirements. R&D departments of OSCs try to identify technology that will either be of interest to their clients or reduce the costs of delivering services. OSCs carefully research and consider prospects and opportunities in the market in order to anticipate changes in demand and meet the future requirements of their clients. Because OSCs aim to customize their R&D to the needs of their clients, they share their R&D plans with IOCs at an early stage to ensure alignment. OSCs aim to balance the necessity of keeping advancements in technology confidential from other OSCs while validating the requirements of their clients. IOCs indicate the potential for growth in the market and help OSCs to anticipate the market trends.

Most OSCs have regular meetings with their customers to understand their needs and challenges. Some OSCs even have an official Client Advisory Board, which consists of representatives of IOCs. This forum provides an opportunity for IOCs to guide the development of new technologies by OSCs.

In addition to influencing the direction of R&D, IOCs also influence the speed of adoption and spread of use of new technologies. A number of OSCs revealed that IOCs direct other OSCs to implement new technologies once they have been developed, as they try to foster competition in the market by ensuring no single OSC has a monopoly on a new technology over an extended period of time.

Moreover, the relationship between IOCs and OSCs in relation to their R&D activities has been changing. Historically the R&D of OSCs has been self-centred and internal with little direct impact from IOCs. However in recent years, R&D in OSCs has been evolving towards an outward-looking model where IOCs have started to take a more active role to the point of actually financing R&D projects at OSCs. The financial participation of IOCs is important for OSCs as it shows IOCs' commitment to their relationship. Similarly, IOCs see value in these partnerships as a mechanism for developing better technologies for challenging oil fields such as ultra-deepwater offshore.

This collaborative approach to R&D mutually benefits both parties and can speed up the commercialization of new technologies. For example, directional drilling, which was developed forty to fifty years ago, has become prevalent only in the last five years. Strategic alliances on R&D projects establish long-term relationships between IOCs and OSCs and hence improve the synchronization between the needs of the market and the development of products. An example of collaborative approach can be seen in the partnership between Total and Halliburton in 2009 to jointly develop a suite of ultra-HPHT measurement and logging while drilling (LWD) sensors. In collaboration with Total, Halliburton worked on the development of the Prometheus suite of LWD tools specifically for the ultra-HPHT environment of some North Sea fields, where harsh conditions challenge the limits of current technology (Dirksen 2009).

Although there are many common characteristics in relationships between IOCs and OSCs, different types of joint ventures and partnerships exist. In some cases, new technology is developed by both the international oil company and the oil services company, with each party taking explicit responsibility for key tasks and outputs. In these instances, the cost of developing the technology is shared between the IOC and OSC. In general if the product is developed together, the IOC maintains exclusive utilization rights for a specified period of time. After this period has lapsed, the OSC has the opportunity to bring the technology to the market and sell it to other clients. For example, the reservoir simulator INTERSECT was created by Schlumberger and Chevron initially. Total came in later, wanting to buy the technology and participate in its further development. Another example which highlights the value of forming joint ventures to develop costly R&D projects is provided by Shell. Shell developed the initial concept of 'expendable tubulars'[8] where expandable tubular material is placed inside the well, which provides more flexibility than standard tubing (Cassidy and Butterfield 2002). Shell secured the patent for the product and then contacted Baker Hughes to jointly develop and commercialize the technology. A company called e^{2}TECH has been established as a 50/50 joint venture of Shell and Baker Hughes in order to develop and market expanded-tube well construction and remediation technology. According to a press release at its time of creation, 'e^{2}TECH will combine Shell's novel expanded tube technology with Baker Hughes' renowned international oilfield service capabilities to reshape the current well construction industry through "in situ steel tube expansion"'. This innovation is aimed at substantially reducing well repair costs and restoring asset integrity while removing many current design limitations (Flaharty 1999). In these circumstances, the IOC has the power to determine the price of the technology and the terms on which it is brought to the commercial market. Time and price advantage is granted to the IOC for technologies developed in partnership.

In other cases, an IOC provides an OSC with the proof of concept for given technology and requests that the OSC develop it according to specifications. Depending on the relationship between the IOC and the OSC, the IOC may agree to finance the development of technology. If the specific technology will be available to other customers on the market, the IOC may negotiate a significant discount or agree to receive royalties, as it generated the initial concept.

Sometimes the IOC develops the technology itself but gives it to an OSC for its commercialization. An example described by an interviewee is in the seismic area. Exxon developed the Surface Slice Application[9] (a seismic data interpretation tool) and allowed Geoquest to further develop and maintain it. In exchange, Geoquest agreed to charge a reduced tariff to Exxon for its services.

There are also cases where an OSC may create a partnership with another OSC in order to use their joint expertise to develop a solution for an IOC. While these partnerships between OSCs are not as common as joint ventures between international oil companies and oil services companies, they do occur. An oil company requested Technip to develop a new technology for subsea integrity and surveillance of flexible pipes. Recognizing the challenge of developing this technology on its own, Technip formed a partnership and signed a global cooperation agreement with Schlumberger. 'By combining Technip's technical and manufacturing knowledge of flexible pipe with Schlumberger surveillance technology, a new generation of intelligent flexible pipe will be created', said Alain Marion, senior VP, Subsea Assets and Technologies, Technip (quoted in Baxter 2009). The partnership between Technip, JGC Engineering and Technical Services and Técnicas Reunidas in Vietnam concerning the development of a refinery is another prime example of cooperation between multiple OSCs. A consortium formed by Tecnicas Reunidas, Technip and JGC Corporation has obtained a $5 billion contract to build an oil refinery in Vietnam (Reuters 2011).

In all the cases above, IOCs give direction to the technological progress either by proposing the idea initially or by developing the technology themselves or in partnership. They may also provide seed funding. For example, Shell formed a venture capital fund to encourage the development of critical technologies.

3.1.4 Close Cooperation vs. Arm's Length Price Relationship

The relationship between IOCs and OSCs is an evolving one. According to industry experts, historically, the relationship was based on close cooperation. Between the 1930s and 1980s, it was centred on technical capabilities, focusing on technology rather than cost. As the oil industry continued to expand in the 1980s and procurement departments in IOCs gained significance, cost considerations increased, and tenders and an arm's length relationship became the norm. Today, the service industry standard for acquiring new business is winning tenders in a competitive market environment. An 'open book approach', where OSCs are completely transparent with international oil companies regarding their balance sheet and profitability, or a 'partnership structure' where IOCs work with the same supplier without considering competitive tenders, are rare.

Although on the surface the relationship is based on arm's length transactions through tenders organized by IOCs, where the lowest price bidder wins the contract, the following paragraphs will attempt to explain why the relationship goes far beyond the simple arm's length price model.

During the tendering process, IOCs are able to influence the outcome of the tender by setting its standards. They might set specific technological requirements, thereby eliminating from the procurement process companies which do not possess the required capabilities. In tenders, bidders are evaluated first based on technology requirements, then on price. Once IOCs confirm the OSCs are prequalified according to technical requirements, they evaluate the submissions based on commercial considerations. Because the oil service industry is highly specialized with a small number of companies offering specialized services, and because IOCs understand the capabilities and competencies of each OSC, IOCs tend to know which company can provide the required services during the preparation of procurement guidelines and tendering rules.

When developing their response to an initial RFP, OSCs consider their technology capabilities, costs and profit requirements and bid at their most competitive price. Although the bid price is a key driver in confirming who is ultimately awarded the contract, an OSC's past experience in the required service is taken into consideration. At similar prices, IOCs prefer to award a contract to an OSC with a successful track record of implementing a given technology or project. Because past experience is important, an OSC needs to improve its track record with new technology. It is often difficult for OSCs with minimal experience in a sector to be successful when tendering for contracts unless they have developed a very niche technology.

In addition to past experience, the established relationships are important. There is a clear trend within IOCs to work with the operators they have established relationships with. IOCs have 'preferred suppliers' in the sense that through working with OSCs, they have a more established relationship with certain companies. Furthermore, an established relationship enables OSCs to understand their customers, and thus to be better positioned to offer the required products and services. OSCs tend to be more aggressive in their bids for repeat business. For example, Técnicas Reunidas worked with Tupras for a long time and developed a solid understanding of the Tupras business model and objectives. This enabled them to be successful in tenders organized by Tupras. Repeat business could result in certain IOCs working with certain OSCs regularly. Anecdotal evidence suggests that BP works regularly with CGGVeritas, Statoil with Aker, Total with Technip and Shell with Schlumberger.

All factors cited above indicate that international oil companies seek a balance between finding strong, reliable and 'aligned' suppliers and ensuring the diversity of suppliers. With this aim, IOCs have been showing a tendency towards 'intelligent arm's length' relationships, where they provide strategic suppliers with information to enable the oil services companies to meet their needs. While international oil companies are not in favour of exclusive partnership agreements with OSCs and prefer competitive tendering, they do prioritize developing relationships with their suppliers. After identifying the market segments by product and services, IOCs aim to have access to and develop relationships with a significant number of suppliers in each segment in order to ensure they have access to the most competitive technology and prices. Recent trends in the industry are increasingly towards the signing of 'Enterprise Frame Agreements' (EFAs) between IOCs and selected OSCs. These agreements do not require exclusivity between the IOC and a given supplier; however, they do highlight the general terms and conditions of the business relationship. The purpose of signing framework agreements is to establish standard prices and service agreements in relation to the procurement of goods and services for a given project. IOCs aim to commoditize the supplied item as much as possible within the limits of the uniqueness of each project.

For example, if an IOC knows that it requires one hundred compressors for a project, it enters into an EFA with an OSC to procure the compressors throughout the duration of the project. The EFA might state that the IOC can purchase up to one hundred compressors within the defined price range between agreed dates. While neither party is obliged to transact, if a business transaction is required, the price and services have been pre-agreed, which facilitates the ease of doing business. For the IOCs, EFAs are preferred to competitive tendering for ordinary items because they ensure standardized designs and bring time and quality benefits. Thanks to the EFAs, IOCs design and narrowly describe the process, the equipment and the service standards with their 'preferred suppliers'. EFAs are also often viewed as the result of IOCs' preference to base their relationships with suppliers on large contracts ($600–700 million per year) in lieu of working with several suppliers on smaller contracts ($10–15 million a year). These global agreements are used for the supply of a specific quantity of goods or services over a given time period on a 'call-off' basis.

Two concrete examples were provided by interviewees from two IOCs. The first example relates to an IOC that signed an agreement with a pressure and well control company for training and skill development. Under the global terms of the contract, the services of the OSC could be requested by any IOC group entity. While the general terms and conditions are set for that specific OSC, the entities have the flexibility to select another training service provider. Another example concerns the agreement between an IOC and a drilling company for shale gas services in the US. According to the agreement, the drilling company must use A-list crew rather than new recruits and offer a price discount to the IOC on publicly quoted rates. The IOC isn't required to use the drilling company exclusively but if it requires the OSC's services, the conditions are set in advance.

All IOCs have signed hundreds of EFAs or agreements of a similar nature including cooperation agreements and call-off agreements, enabling them to establish contractual relationships with multiple suppliers. IOCs and OSCs work together to determine and agree which goods and services can be procured with EFA contracts. According to an example given by an interviewee, an IOC requested a specific flow control mechanism and the OSC began producing the model exclusively for that particular IOC. However the specific model was 30% more expensive than a standard model which has the same functionality. The OSC advised the IOC that a standard model could be used by the IOC under an EFA contract, which led to cost savings and efficiency for the IOC. Following the agreement of the IOC, an EFA has been signed for the standard model.

Finally, the tendency towards 'intelligent arm's length' relationships manifests in regular coordination between IOCs and OSCs. Parties have periodic service reviews where they focus on ways to improve the relationship and provision of services. Most IOCs have annual meetings with OSCs to discuss new technologies, and new commercial and contractual issues. IOCs and OSCs increasingly coordinate their activities in order to better understand the requirements and objectives of each party. This is particularly important because the next generation of oil will be located in areas that pose complex challenges.

3.1.5 Procurement within International Oil Companies

Quality, time and cost are fundamental elements of every supply chain. International oil companies require services to be of good quality, punctual and at the agreed price. The profitability of an oil field has been primarily determined by the amount of oil found and the current price of oil. Although the cost of OSCs has been the main expenditure of IOCs, until recently it had secondary importance for the profitability of an oil field. Historically, purchasing was done at a local level in a decentralized manner. However, in recent years, the cost of services has skyrocketed due to the increase in the cost of materials and the lack of spare capacity within the oil services sector. This is evidenced in the doubling of the IHS CERA Upstream Capital Costs Index between 2005 and 2008 (IHS 2015; see Figure 3.1 for more details).

The increasing cost of oil services has boosted the importance of cost controlling and procurement within IOCs in the last decade. Procurement activity and supply chain management have become more centralized and professionalized. Although IOCs differ, the general tendency is towards centralization of procurement at a global level. Exxon and Chevron are known to have centralized procurement whereas other IOCs such as Shell, BP, Total and ENI have only started to centralize their procurement in recent years. By doing so, they aim to improve the procurement process, creating transparency between branches, headquarters and suppliers.

In general, centralized procurement puts greater pressure on suppliers. The aim has also been to standardize at least the large items (instead of customizing each single well), even though standardization is challenging because of the complexity of projects. Risks are reduced when there is a central structure when buying standardized services. Procurement is therefore

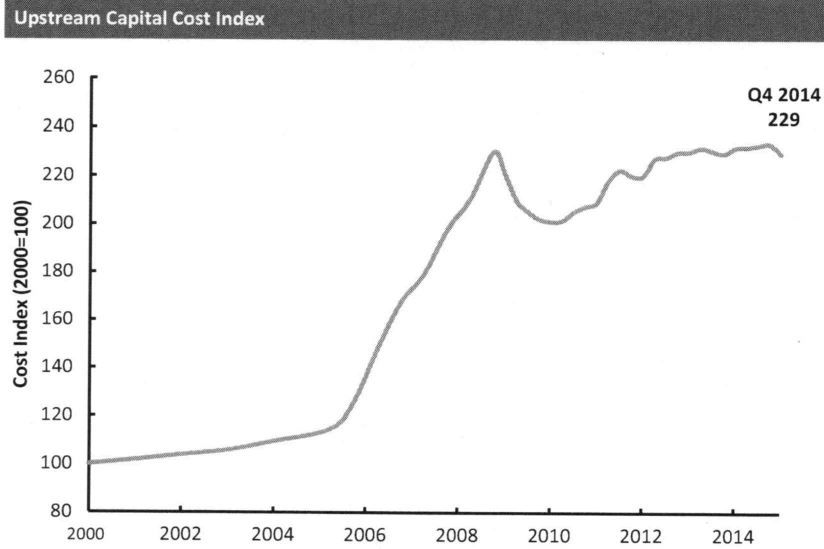

Figure 3.1 IHS CERA Upstream Capital Cost Index

Source: IHS Energy (IHS 2015).

Note: The use of this content was authorized in advance by IHS Inc. Any further use or redistribution of this content is strictly prohibited without written permission by IHS Inc. All rights reserved.

done worldwide in a centralized manner from the same central entity. The central body takes charge of procurement and makes sure that the procured items arrive on time and meet necessary quality standards.

Procurement departments aim to better assess the needs of the IOC and systematize the exchange of opinion with service suppliers. Points of satisfaction and dissatisfaction are discussed during supplier quality review meetings. There are also forums for the operators (IOCs) to discuss industry practices and contractor experiences with each other.

Within the procurement department, a separate procurement strategy is laid out for each big project, segment and service. For large projects, IOCs design a 'statement of requirement', which is similar to a scope of work. The document includes specifications, planning, budget, priorities and constraints. Furthermore, IOCs investigate market availability of the required services as specified in the statement of requirement. If the products or services are available on the market, then the IOC defines a procurement strategy. In a typical procedure, the IOC organizes a tender where bidders are prequalified, the target price is determined and the criteria for evaluation (technical, legal, HSE, commercial, financial, etc.) are fixed. In general, there is no policy of blacklisting but the IOC takes into consideration feasibility and the pros and cons of working with each supplier before selecting a preferred supplier accordingly. For example, the supplier might be weak in project planning and management but possess first-class capability in gathering subsurface data. If the IOC decides to proceed with that supplier, it will take the OSC's weak points into consideration in order to compensate for them. The IOC's in-depth knowledge on capabilities of OSCs is an important aspect of system integration. There are some contractors that are listed to be used for 'less challenging projects'. Contractors must provide a satisfactory service for each project in order to avoid being added to negative listings.

If the technology requested is not available on the market, international oil companies inform the suppliers about the need and ask them to propose corresponding solutions. The IOC analyses the proposed solutions and selects the preferred option. In these cases, IOCs don't specify the solution; they provide OSCs with information regarding their requirements, and suppliers provide solutions to address the issue.

The unique nature of each project requires a different strategy and impedes international oil companies from having serial or standard project management contracts with OSCs. Despite the nonexistence of serial contracts, an IOC's relationship with its suppliers is typically long-term and well established. For example, IOCs and OSCs meet once a year at the CEO level. At the project level, they meet regularly to exchange feedback. Large IOCs and OSCs have been working in partnership for much of their history. For example, Shell and Schlumberger have been working together for more than fifty years.

With procurement activities gaining importance, IOCs have changed their contractual strategies and developed different approaches for procurement. In general, IOCs set a group procurement policy applied by the procurement department. Each branch (LNG, gas, exploration) and business unit within the group has a vertical connection to the procurement division and abides by common procurement procedures. Depending on the size and nature of the goods required, it may be possible for a local department to procure the item, but the central procurement department needs to oversee purchases over a certain threshold (say, over €50 million). For example, for well testing, if data-gathering services are required, contracts will be set up by the well services teams. If the contract is over a certain amount, they need to get the procurement department involved. Procurement teams are also involved for long-lead items or for items that are required in large quantities, such as compressors and pipes.

International oil companies establish specific procurement targets for each year (e.g., 80% of purchases will be done through the procurement line). Certain IOCs go through the tenders and record bids with the lowest and highest price in addition to the price of the selected bid. The difference between the highest price and the contract selected is recorded as 'saving for the company'. Then a process takes place every month where all spending is collected and analysed to identify how effective these savings initiatives were.

Certain IOCs have developed divisions for market intelligence within their procurement departments. The 'market intelligence' divisions understand the market, the prices of the feedstock (such as steel) and the prices of equipment that OSCs use (such as compressors). IOCs try to build a better understanding of the costs of OSCs and their profit margins. IOCs are careful to check the suppliers' costs and profits in order to make sure the supplier continues to invest while not charging excessive fees to the IOC. The main objective is not to squeeze the margins but to better understand the costs and profit margin of oil services companies. International oil companies aim to increase their purchasing power in the marketplace by improving their assessment of costs and managing global procurement from one entity. This is an example of profound system integration from the top to the bottom of the supply chain. International oil companies are 'upstreaming' with the aim of controlling costs.

Several IOCs have also developed category management systems. The category managers follow up large accounts with the objective of developing short-, mid- and long-term strategies for procurement. Category managers analyse the market and market participants to determine negotiation leverages and a procurement strategy for each segment. They also codify and create norms for each category of services and products. Category managers question whether the IOC should go ahead with premium suppliers or nominate other suppliers. For instance, if 40% of drilling activity is in deepwater, category managers question whether the IOC should purchase a deepwater driller or contract deepwater drilling to an OSC. Another example can be given for the subsea segment. International oil companies use horizontal trees or vertical trees in different fields. The category manager assesses which approach is better and whether there should be a standardization of the equipment. They also try to maintain the transparency of the market by ensuring competition for each segment, which means following thousands of suppliers. An IOC has stated that it follows 3,500 suppliers. Category managers make a list of 'preferred suppliers' according to the prices and the technical expertise of suppliers and sign cooperation agreements or EFAs where necessary. The IOC in question has over 800 cooperation agreements. Another IOC stated that it defines a category strategy for each sector, subsector, technology and item (such as subsurface technologies, rigs, etc.) and further defines the type of supply and the preferred approach within that category.

One of the main factors that has changed in procurement departments is that IT developments (such as SAP enterprise resource planning software) have allowed oil companies to track their worldwide purchases and compare costs. Systems like SAP have enabled IOCs to compare spending on each item and have become a key tool for procurement divisions.

Some of the IOCs acknowledge that procurement and supply chain management are different and that supply chain management can generate longer term savings. IOCs increasingly aim to focus on strategic purchasing, sectors of spending (drilling, etc.) and developing relationships with key suppliers. Purchasing conventions are organized every year where local divisions of IOCs discuss procurement issues.

The creation and empowerment of central procurement divisions has a huge impact on the way that business is done today between IOCs and OSCs. Certain OSCs believe that the rise of procurement departments has caused setbacks in their relationship with IOCs. Increasingly, OSCs transact with procurement departments rather than the end-users of the service or equipment within the IOC. OSCs complain that it is more difficult to prove the value they add to the process, and to differentiate their goods and services to procurement department personnel. When end-users of the service and technical experts (for example experts in the well design division of the IOC) of both IOCs and OSCs interact, it particularly benefits those OSCs with significant technical capabilities.

In the words of an interviewee working for an OSC:

> An end-user appreciates the value of what we [OSCs] do. A procurement department does not. After technically qualifying companies at the minimum common denominator, procurement departments see no difference in services. Although procurement personnel have a minimum amount of technical qualifications, they are unable to differentiate between specific services and capabilities. There has to be a better way of qualifying benefits that cannot be expressed on an Excel sheet. For instance, companies should get penalized if they cannot provide a certain level of solutions. This penalty can be translated into money and included with the bidding price. This way, the good technological solutions would be easily highlighted in the bidding process. Often good technology is slightly more expensive; the price difference is difficult to explain to procurement departments. Prequalification during the tenders indicates a minimum technological requirement but providing the most technologically advanced solution does not affect the outcome. Technological progress is not distinguished during the tender. We propose the minimum technology required, win the tender and then discuss more advanced technology. If we went ahead with advanced technology with its associated price tag at the beginning, we would lose the tender. In addition, when procurement is involved, OSCs aim to win with a lower price but then try to increase the price by offering alternative options. This causes an unproductive cycle where OSCs try to increase the fees and IOCs try to keep them low. Procurement departments cannot master the total cost of a service. It goes beyond the price per meter or per hour. Elements such as accidents, wrong measurements and the

> cost of poor quality are not calculated. For example, Shell made a wrong estimation of oil reserves due to an incorrect measurement. No project post-mortem was done to evaluate the total costs. We need to develop some sort of multiplier that can be applied to pricing as a way to estimate potential total costs. Procurement departments do not see that.

OSCs believe that as the supply chain of an organization develops, the end-users become increasingly distanced from service providers. When contracting was done by end-users within the IOC, they possessed an in-depth knowledge of the advantages of an advanced technology and the details of the required service. According to OSCs, all components such as price, technological solution, HSE record and local content should be given more weight during tendering.

In terms of procurement strategy, an interviewee working for an IOC confirmed that if services require a specific technology then the IOC will choose an OSC which can offer it at premium quality. In complex projects, IOCs prefer to do business with high-end contractors, which are, in general, international suppliers. The large OSCs such as Schlumberger, Halliburton, Transocean, Baker Hughes, Weatherford, Technip and Saipem come at the top of the list of main suppliers for most IOCs. If the service is a standard one that can be performed by any OSC, then an IOC will select the least-expensive option on offer. At the end, IOCs' ultimate aim is to find the best value.

In order to better meet the needs of IOCs, OSCs must have strong supply chain management too. Oil services companies try to form strategic partnerships with their subcontractors (i.e., owners of raw materials) while also aiming to diversify supply. Most OSCs have a central procurement that serves all internal departments. The intention is to purchase required items under one contract. Schlumberger is a prime example of an OSC that can effectively manage its supply chain. Schlumberger does not outsource services but rather outsources certain manufacturing tasks while doing the final assembly and testing in-house. Suppliers are of great importance to large OSCs such as Schlumberger, and they therefore manage these relationships very carefully. Schlumberger has a very strong supply chain and procurement organization. Schlumberger Supply Chain Services provides Schlumberger entities with supplier management, strategic sourcing and logistics and inventory management. Schlumberger selects preferred suppliers among those that work in a professional, ethical, competitive and cost-effective manner consistent with Schlumberger policies, procedures and business objectives (Schlumberger 2012a).

Procurement divisions in other OSCs are also becoming increasingly more sophisticated. For instance an interviewee from an OSC said that in 2009, the OSC hosted a conference with the management of 150 of its suppliers to discuss ways to improve their working relationships and overcome challenges facing their businesses. During the conference, it assigned an employee to work directly with each of its suppliers. The employees effectively became the sponsor and account manager for the supplier and were tasked with developing a strong understanding of the supplier's business model.

As detailed above, the relationship between IOCs and OSCs is complex and shaped by multiple factors. The next section will investigate two additional dynamics that affect this relationship, namely the presence of national oil companies and joint venture structures.

3.1.6 Relationship under the Shadow of NOCs

Oil has a narrow geographic distribution of supply. International oil companies work on fields owned by host countries that have their own national oil companies. NOCs vary widely in size, technical capability, capital structure and geographic footprint. Some NOCs such as Statoil and Petrochina are, like IOCs, publicly listed companies. Others such as Kuwait Petroleum or Nigerian National Petroleum Corporation are state companies which carry the characteristics of public administration. Certain NOCs such as Nigerian National Petroleum Corporation and KazMunayGas are considered less advanced by industry experts when compared to others like Statoil, Petrobras and Saudi Aramco, all of which possess advanced technologies. As an example, Saudi Aramco is the largest oil corporation in the world with the largest proven oil reserves and has its origins in Seven Sisters[10] (Hermann et al. 2010). It has been ranked as number one oil company in PIW rankings for twenty-two consecutive years (Saudi Aramco 2010). According to interviewees, in the 1980s and 1990s Saudi Aramco did not have the same technical capability as the majority of IOCs. However, after decades of investment and working with its suppliers, Saudi Aramco has developed first-class technology, significant expertise and technical capability to adapt new proven technologies very quickly. Petrobras is cited as another example of a sophisticated NOC, being at the forefront of technology in the subsea and deepwater. While most NOCs predominantly operate in their own country, some such as Petrochina have significantly expanded to overseas markets, acting like an IOC.

Depending on the capabilities of its NOC, the host country takes the decision whether to rely on its own NOC or contract an IOC to develop an oil field. Host countries tend to use IOCs for big fields that require large-scale financing. In the case of smaller and uncomplicated fields, they bypass IOCs and ask their NOC to work directly with OSCs. To be awarded a contract, IOCs need to prove their added value regarding financing, technical capability and operational efficiency. This is particularly challenging when oil prices are high. In this case, host countries would have more funds to invest in their own fields and may consider loss of operational efficiency less important. However, IOCs still have a valid point to make as there are examples where operational efficiency of an oilfield dropped significantly under the management of NOCs. For example, in Mexico, oil production of its NOC PEMEX declined by 25% from 3.3 Mbbl/day to 2.5 Mbbl/day from 2006 to 2011 (Daly 2012).

There is a significant amount of debate regarding the role of IOCs in the future. Accordingly, if OSCs offer integrated services and NOCs acquire all the technological competences necessary to be able to supervise their work, they may bypass the international oil companies that exist in between NOCs and OSCs. Certain sophisticated NOCs (such as Petrobras) already act like IOCs and have developed their own supply chain. Today, Saudi Aramco and Russian oil companies use OSCs directly.

As explained above, NOCs are more likely to develop oil fields with OSCs if the conditions for oil extraction are straightforward, but they seek the help of IOCs when conditions are complicated. For example, in the Shtokman natural gas field, which experiences extreme Arctic conditions, Gazprom sought the participation of IOCs to assist in the development of the field. The situation varies for onshore and offshore fields as well. For onshore fields, NOCs tend to integrate the activities and manage the supply chain themselves. For offshore drilling and production, NOCs tend to choose IOCs as supply chain integrators. Major differences between onshore and offshore operations lie in the need for significant upfront capital expenditure for offshore development, and the significant technical expertise required to extract oil offshore. When the field requires a drilling depth up to 3 kilometres, NOCs require IOCs' access to large capital and advanced technology.

When they need to work together, NOCs and IOCs typically form partnerships under joint venture structures and work under production sharing agreements (PSAs). In PSAs, IOCs recover the cost of exploration and development with the oil produced. Therefore, NOCs impose rules for tenders and may determine the choice of OSCs because the money being spent is the resource of the host country. In several countries, host countries impose systematic tenders. NOCs may also insist on the use of local services companies via 'local content requirement' clauses.

OSCs are agnostic with regard to doing business with NOCs or IOCs. They work directly with NOCs in several fields and offer integrated project management services for NOCs. In the words of an interviewee: 'Some OSCs are very large that own highly significant technologies for field development. It is seen as critical for the development of the oil field by some NOCs. For example, Schlumberger is offering integrated services to Pemex in Mexico and it is considered crucial by Pemex'.

For OSCs, IOCs provide fewer but larger opportunities compared to NOCs. While IOCs have access to less than 10% of the reserves and produce around 15% of global oil, they manage 30%[11] of 'new' oil field development globally. OSCs prefer to be contracted for one large project by an IOC than for several smaller projects. Certain OSCs acknowledged that they prefer IOCs as an interface because having a direct relationship with

NOCs can be less reliable and more 'rocky'. OSCs encourage NOCs to utilize IOCs due to their experience and technology. OSCs also state that certain NOCs lack structure in procurement and contract management. For instance, from tendering to reward, contracts can take up to two years to be implemented with NOCs whereas international oil companies have a much shorter procurement process. The language barrier with NOCs is also cited by some OSCs as the simple reason for their preference for working with IOCs. Several OSCs see IOCs as long-term partners in comparison to NOC arrangements, which tend to be one-off deals.

There is a regular debate in the industry regarding the power shift between NOCs and IOCs and how this will have an impact on the relationship between IOCs and OSCs. As stated, certain NOCs are becoming as sophisticated as IOCs and are contracting OSCs directly. Furthermore, they are increasingly attracting high-calibre talent. Engineers who used to work for IOCs move to NOCs and bring along their expertise.

These developments have caused a decrease in the number of PSAs and an increase in technical services contracts between NOCs and IOCs. IOCs are being forced to develop a service-oriented approach when dealing with NOCs. For example, in Iraq, recently IOCs had to agree to sign service agreements, and certain OSCs mentioned that in the future there might be tenders for which they will have to compete with IOCs to provide the same services to NOCs. The fact that IOCs are entering into a greater number of service contracts will redefine the relationship between IOCs and OSCs.

3.1.7 Relationship within the Context of Joint Ventures

As mentioned previously, NOCs and IOCs typically work together under JV structures. This is a contrast to some other industries such as the aircraft industry where Boeing and Airbus are standalone companies that assemble products in their own facilities. Today most large oil fields are explored and developed by joint ventures where several international oil companies and national oil companies form partnerships. One of the IOCs is usually chosen as the operator of the field and manages the overall process and procurement. The operator heads the 'Operation Committee', selects the OSCs and submits the choice to its partners. The allocation of an oil services contract is decided together with JV partners. When an international oil company is known for its experience in a given area, operators or other JV partners would expect that international oil company to have more input on decisions related to that area. Partners approve the choice of the OSCs as well as the costs involved. In JVs, costs are paid upfront and all expenses are submitted to annual audits at the end of the year. For smaller amounts up to a certain limit, the operator has the sole authority to make decisions. Above the agreed amount, tender selections are submitted to 'bid committees'. Because the choice of OSC is made with JV partners, the progress of services is regularly communicated to the partners. In recent years, the front-end loading concept has gained prominence in the industry. In front-end loading, IOCs do extensive analysis during the E&P appraisal phase and most decisions regarding supply chain are decided up front following these analyses.

There are different configurations and competitive pressures in the joint venture structure. The type of investment and relationships with OSCs are related. An IOC's positioning is different when working alone, in a JV or with an NOC. The IOC might have more flexibility in a joint venture where NOCs are not involved, as NOCs might impose additional conditions such as local content requirements.

3.2 IMPLICATIONS OF THE RELATIONSHIP

The significance of the relationship between IOCs and OSCs is often understated by industry observers. By contrast, the importance of the relationship between IOCs and national governments or NOCs is highly recognized. The constant need to access new reserves in order to grow their business makes IOCs' connections with resource holders of crucial importance. On the face of it, the link between IOCs and OSCs seems less significant. However, it is equally vital. A wrong choice of OSC or work done poorly by an OSC will impact an IOC's business and its relationship with governments. Furthermore, access to reserves requires technology provided by OSCs and efficient teamwork from both OSCs and IOCs.

The following section will analyse the importance of this relationship by looking at its financial and technological significance. It will also highlight how this relationship has shaped the oil services industry structure and how collaboration and rivalry are maintained in fine equilibrium. Finally, the section will provide us with the unfortunate example of the BP Macondo accident, where the problems in the relationship (among other factors) caused substantial negative consequences.

3.2.1 Financial Significance

OSCs provide assets, technology and staff across all segments of the oil supply chain and generate revenue by services provided to clients including IOCs, NOCs and independents. The revenue of OSCs forms the most important capital expenditure of IOCs. The expenditure of IOCs on OSCs consists of *exploration costs* (seismic and exploration and appraisal drilling), *developments costs* (front-end engineering and design, procurement of equipment, construction of facilities, drilling, vessel/rig purchase, engineering and project management) and *operating costs* (day-to-day operating expenses such as cost of fuel, aircraft, catering on the rig, transportation and other logistics and daily maintenance) (Hermann et al. 2010).

The capital expenditure of IOCs, most of which is paid to OSCs, is substantial and demonstrates the significance of the relationship between IOCs and OSCs. For instance, Shell and Exxon spent $26 billion and $36 billion respectively as capital expenditure in 2011 (ExxonMobil 2011; Shell 2011a; see Table 2.4 for more information on expenditure of IOCs). The large procurement budgets are likely to increase even further because the industry is moving into deeper and more challenging fields. The IHS CERA Cost Index highlights that between 2004 and 2008 CAPEX costs in the oil and gas industry more than doubled. The IHS CERA Upstream Capital Costs Index registered a record high in 2008.

The costs are very substantial even at the project level, where the contracts between service providers and IOCs often involve large sums. For example, a single floating production, storage and offloading (FPSO) vessel at the centre of project Girassol in Angola accounted for $920 million out of the $2.8 billion total project costs (Total 2003).

3.2.2 Technical Challenges

The decline of easy-to-access oil fields has increased the need for IOCs to develop new resources in remote and hostile environments. IOCs maintain competitive advantage and convince host governments to access reserves by leveraging their technological competence. OSCs are technology providers to IOCs. It is essential for IOCs to remain experts in leading and coordinating different technologies provided by OSCs. Therefore, the relationship between IOCs and OSCs is critical for IOCs to overcome technological difficulties and hence to unlock the oil and gas reserves in challenging fields. A close collaboration on the development as well as the application of technology is required to defeat technical challenges. Technical divisions of IOCs are therefore constantly in contact with OSCs. Increasingly, technology is also advanced through technology joint ventures between IOCs and OSCs.

3.2.3 Competition versus Cooperation

As explained in previous sections, the business models of IOCs and OSCs differ. IOCs finance oil projects and integrate different technologies provided by OSCs. In general, OSCs specialize in particular technologies and offer these services to IOCs. A few large OSCs such as Schlumberger and Halliburton are exceptions as they provide a broad range of services. Despite the contrasting business models, the question of competition between OSCs and international oil companies has been posed since the '90s.

In the past, it was questioned whether OSCs would eventually challenge IOCs. Certain NOCs (such as Petrobras and Saudi Aramco) have become as sophisticated as IOCs and have started contracting OSCs directly, especially for technically simple onshore fields. NOCs either combine multiple services provided by different OSCs or hire a large OSC such as Schlumberger, which offers integrated services. The competition challenge to IOCs has been particularly valid in the case of Schlumberger, and to a certain extent Halliburton, because both companies offer a complete range of services as opposed to other service companies such as Transocean or Technip, which are specialized in specific technologies such as drilling for the former and subsea technologies for the latter. Today, OSCs like Schlumberger and Halliburton propose 'integrated project management' to NOCs. However, despite 'payment out of production' being maintained in certain contracts, this is limited to 'a fixed price paid in the form of oil' and does not extend to 'a share of oil found in the reserve'. In its 2011 annual report, Schlumberger states that 'projects may be fixed price in nature, contain penalties for non-performance and may also offer opportunities for bonus payments where performance exceeds agreed targets. In no circumstances does Schlumberger take any stake in the ownership of oil or gas reserves' (Schlumberger 2011, 22). In certain production enhancement contracts that aim to increase oil production mainly in depleting fields, OSCs agree with their client that they will share some of the risk and be paid relative to the level of additional production. In these contracts, strict conditions are attached to payments linked to oil production. OSCs cannot be equity owners and are therefore prevented from having a percentage share of the oil reserves. For example, Saudi Aramco awarded its first integrated project management contract to Halliburton, where Halliburton's payment will be linked to the quantity of increased production. Although the payment is linked to production, it is still a fixed-price contract, which is different from the contracts awarded to international oil companies where IOCs take a share in oil equity.

An additional example is offered by Schlumberger, which works on mature fields with a declining production curve and takes a share of additional production it increases. The interviewees mentioned that this type of work does not pose a challenge to IOCs because Schlumberger takes the risk of increasing oil production but does not, as IOCs do, undertake the risk of finding oil. In addition, these types of contracts are for onshore depleted fields and not for new prospective fields to which IOCs seek access. Last, Schlumberger is not involved in oil reserve participation but its payment is linked to additional amount of oil produced. A situation where OSCs share a percentage of oil reserves and hence take 'the oil risk' would be at odds with the business model of IOCs.

Interviewees from OSCs rejected the idea of competition with IOCs. They highlighted that rivalry with their own customer would impact upon their conventional business and cause a conflict of interest. It was their stated aim to help IOCs be successful by improving quality and developing technology.

A few IOCs accepted that the challenge posed by integrated OSCs such as Schlumberger and Halliburton is a real concern, with several interviewees from IOCs revealing that they believe there is a need to evolve their strategy regarding Schlumberger because it is a well-integrated company which is financially comparable to certain IOCs and employable directly by NOCs.

However, other IOCs disagreed that a challenge was being posed by OSCs, stating that OSCs would be a threat if they took away opportunities for IOCs and arguing that currently business domains are well separated. Although they increasingly develop technologies together, business expertise remains separate. Accordingly, rivalry takes place over the sharing of economic rent but not in terms of business activity.

It is also worth noting that some interviewees from OSCs claimed that competition could also occur in the other direction, with IOCs accepting structures similar to OSC service contracts. Accordingly, the standard model of 'service fee for OSCs' and 'equity share for IOCs' shows signs of change due to recent transformations in the industry.

Traditionally, IOCs have worked under two distinct petroleum regimes. The first is the 'production sharing agreement' (PSA) regime in which the host government retains ownership of petroleum resources and oil companies receive a share of production or revenues from the sale of oil. Second, there is the 'concession system' where the government assigns the right to explore and develop its petroleum resources to the IOC in return for a share of the proceeds, or a production-based royalty. In most countries,[12] the government owns all mineral resources but the title is transferred to the IOC when oil is produced. In return, oil companies pay royalties and taxes to the government (Farnejad 2006).

In recent years IOCs have been 'requested' to accept fee structures similar to OSCs. For example, in Iran IOCs had to accept 'buy-back' contracts[13] (Menas 2012). These are essentially 'risk-service contracts' where the contractor funds the investment, takes the risk and receives remuneration via an allocated production share. Under this model, the IOC transfers the operation of the field to the NOC after a set number of years and never gains equity rights in crude oil (Farnejad 2006). Within the framework of the petroleum sector, a buy-back contract is a mere 'service contract' with a limited life span and limited risks and rewards. Similarly, in Iraq, instead of industry-standard production-sharing contracts, IOCs accepted service contracts paying them a fixed fee for the amount of oil produced. For example, the joint venture between Shell and Petronas won the right to develop the Majnoon oil field with a twenty-year service contract under which they will receive a fixed fee of $1.39 per barrel produced (BBC 2009).

Based on what IOCs agreed to in Iraq recently, certain OSCs mentioned the future possibility of IOCs and OSCs participating in the same tender and bidding against each other. Therefore, it is possible to see that IOCs and OSCs may compete to provide the same services.

Certain OSCs stated their apprehension about the model of IOCs coming closer to their own 'service model'. They highlighted that the IOCs' strategy is difficult to read and their future hard to guess. Accordingly, the future is more uncertain for IOCs than it is for OSCs. This is reflected in their share price performance over the past years.

Table 3.2 below shows the comparison of share price performance of selected IOCs and OSCs. The data show that in terms of share price, IOCs[14] mostly outperformed large OSCs, such as Schlumberger, Halliburton and Transocean,[15] between 1990 and 2000. However, an overall comparison confirms that most OSCs outperformed IOCs over the past twenty-five years.

The question of competition versus cooperation is currently only relevant in relation to a limited number of OSCs (mainly Schlumberger and Halliburton) and IOCs. However, as detailed above, the issue is evaluated differently by several interviewees. Current and potential future competition

Table 3.2 Stock Indices of Selected IOCs and OSCs

Share Prices per Company	Currency	Base Year	2000	2015
British Petroleum (BP)	GBX	169	596	411
Royal Dutch Shell (SHELL)	EURO	8.42	29.73	27.66
Total (FP FP)	EURO	5.40	32.69	42.52
Schlumberger (SLB)	USD	11.02	27.47	85.41
Halliburton (HAL)	USD	10.22	19.16	39.33
Transocean (RIG)	USD	15.81	46.90	18.33
Weatherford International (WFT)	USD	1.08	6.01	11.45
Cameron International (CAM)	USD	2.19	11.19	49.95
Indexed Share Prices	**Base Year**	**Base Year**	**2000 vs. Base**	**2015 vs. Base**
British Petroleum (BP)	1990	100	353	243
Royal Dutch Shell (SHELL)	1990	100	353	328
Total (FP FP)	1990	100	606	788
Schlumberger (SLB)	1990	100	249	775
Halliburton (HAL)	1990	100	188	385
Transocean (RIG)	1993	100	297	116
Weatherford International (WFT)	1990	100	558	1061
Cameron International (CAM)	1995	100	511	2283

Source: Compiled by the author using Bloomberg data (as of May 22, 2015). Used with permission of Bloomberg L.P. Copyright© 2015. All rights reserved.

Note: CAM: Cameron, RIG: Transocean, HAL: Halliburton, WFT: Weatherford International, SLB: Schlumberger as quoted in the US stock exchange; BP: British Petroleum as quoted in LSE; FP FP: Total as quoted in French Stock Exchange and SHELL: Royal Dutch Shell as quoted in the Dutch Stock Exchange. All share prices are indexed to 100 starting from the date of their availability: 1993 for Transocean, 1995 for Cameron and 1990 for all others. Base year is indexed to 100. All share prices are as of the first available day of the year in question.

would require further investigation. However, the competition is worthy of note because it is hard to see suppliers being able to compete with system integrators in other sectors such as aircraft or automobile.

3.3 IMPLICATIONS OF THE CHANGES FOR THE VALUE CHAIN

3.3.1 Strategic Implications

Previous sections described how changes in oil prices and in the size of projects impacted the OSC market structure. This section will specify strategic implications of the changes in the supply chain and in the industrial structure of OSCs.

As system integrators, IOCs have an important role in shaping the industrial structure of OSCs. As expressed by an interviewee, 'IOCs might not be directly controlling industrial restructuring, but they cause it'. For example, cost-cutting pressure from IOCs in the '80s and '90s due to low oil prices was one of the catalysts for consolidation in the services sector. Through the 2000s when exploration and development activities peaked due to high oil prices, IOCs found themselves in a difficult position negotiating with a limited number of OSCs already working at full capacity. Consequently, IOCs sought to encourage competition among service providers in order to control costs. They favoured the emergence of new players and aimed to distribute work among multiple service providers. An interviewee cited that when a new technology is developed by an OSC, IOCs seek to diffuse it to other OSCs to avoid the market being monopolized by one supplier. According to OSCs, IOCs monitor the share of particular service providers in their portfolio and will choose a different supplier if the share is considered to be too high. Alongside their own portfolio, they also observe the overall market share of each provider and promote other providers to ensure competition. For example, an interviewee mentioned that BP supported CGGVeritas to increase competition against Schlumberger—which became a dominant player in the first place due to its close relationship with Shell, one interviewee suggested, although others cited that Schlumberger's powerful position among OSCs can be attributed to the capabilities it has built independent of Shell.

Several interviewees from IOCs emphasized that IOCs need to ensure that they attain the right balance between maintaining the financial health of the OSCs, promoting competition and transferring an appropriate level of risk to OSCs. Accordingly, the two areas to improve are the balance of risk between IOCs and OSCs and consolidation in the service sector. As system integrators, IOCs shape the industrial structure by continuously adjusting the balance of risk between themselves and OSCs and ensuring competition among OSCs.

3.3.2 Operational & Managerial Implications: Example of the BP Accident

A tragic accident in 2010 illustrated the extent of system integration, the dimensions of the extended company and the operational and managerial implications of the changes in the petroleum supply chain.

On April 20, 2010, one of the biggest accidents in oil industry history occurred in the Gulf of Mexico. A well control event (wellhead blowout[16]) allowed hydrocarbons to escape from the Macondo well onto Transocean's Deepwater Horizon rig. This resulted in an explosion and fire, which eventually caused the rig to sink. The accident took place when the rig was drilling a well at a depth of 1,525 metres approximately forty miles from land, on what should have been one of its last days on the well[17] (Financial Times 2010a, 2010b). The oil continued to flow from the reservoir to the wellbore and blowout preventer (BOP)[18] for eighty-seven days, causing the largest accidental marine oil spill in the history of the oil industry[19] (Crooks and McNulty 2010; Hoyos 2010). The leak was eventually stopped; first temporarily by capping the wellhead on July 15, 2010, and then permanently by the completion of a relief well that intercepted the leaking well, pumping in mud and cement on September 19, 2010. Although the circumstances that the Deepwater Horizon rig was working under were exceptionally difficult, the situation was not unprecedented. In 2009, the same Deepwater Horizon rig had drilled the deepest well ever completed, the Tiber oil field, which had a vertical depth of 10,683 metres (deeper than the height of Mount Everest).

From the system integration point of view, the BP disaster is markedly different from other accidents such as the Texas refinery explosion[20] or oil spills such as Exxon Valdez. No direct contractor involvement with the accident is apparent in BP's Texas refinery accident. Similarly, in the Exxon Valdez oil spill, no third party was involved because the tanker belonged to Exxon Shipping Company. In comparison, the BP accident involved multiple contractors and demonstrates the problems with system integration.

The BP accident was the result of several failures rather than a single action or inaction.[21] BP's investigation team concluded that a complex and interlinked series of mechanical failures, human misjudgement, engineering design flaws, operational errors and gaps in communication between teams came together to allow the accident and its subsequent escalation (BP 2010b).

BP faced one of the biggest corporate crises ever, experiencing disastrous financial as well as public relations consequences. In 2010 at the worst moment of the crisis, BP suffered a loss of over 50% in market value and reported second quarter losses of $17 billion, its first loss in eighteen years. The Gulf of Mexico oil leak left the company with costs of over $32.2 billion including the $20 billion compensation fund it set up. BP initiated the largest response by a single company to an accident, deploying nearly 47,000 people at the peak of the crisis with over 1,000 scientists and experts gathering together to find a solution to stop the oil spill (Hayward 2010a). Simultaneous operations of up to nineteen principal vessels within a 500-meter radius of the wellhead, as well as an additional forty to fifty vessels within a one-mile radius, showed the unprecedented scale of the response to the accident (BP 2010a).

The accident can be analysed from different perspectives and will certainly have repercussions on oil industry practices. It will have an impact on regulation, lead to additional security systems and governance and increase the time and cost of oil projects. Christophe de Margerie, the chief executive of Total, said that new oil exploration in the Gulf of Mexico was likely to take 20% more time and cost 20% more as a result of the new regulations (Christophe de Margerie quoted in Crooks 2010a). The aim of this section is to analyse the accident by looking at the system integration functions of an IOC, the results of an oversight in these functions and the liabilities attached to it.

As is usually the case on oil fields, several companies were involved in the Macondo well. BP Exploration & Production Inc. was the lease operator of Mississippi Canyon Block 252, which contains the Macondo well. The Macondo prospect is owned jointly by BP (65%), Anadarko Petroleum Corporation (25%) and MOE Offshore, a unit of Mitsui (10%). The rig, namely Deepwater Horizon, is a semi-submersible drilling unit owned and operated by Transocean under contract to BP. Cementing and mud logging services for the well were provided by Halliburton and the drilling fluid was supplied by MI Swaco (a unit of Schlumberger). BOP was manufactured by Cameron and the float collar by Weatherford International. In addition, Schlumberger performed wireline services but left the rig on the day of the accident. Schlumberger was also contracted to do a cement bond log test but BP decided to skip the test. BP as the operator was responsible for the integration of different services and final decisions related to these services, to wit BP defined the drilling programme, designed the drilling unit and chose the type of BOP to be used. As is usually the case on the platform, the decisions were taken by BP but executed by service companies such as Transocean and Halliburton.

Following the accident, several theories have been suggested regarding its 'fundamental' cause. The industry questioned whether the accident happened as a result of too much outsourcing, or whether the problem was one of integration. Several industry experts concluded that risk would not have been reduced if services had remained in-house. However, it was mentioned that the risk would have been lessened if IOCs had strengthened their expertise and been better able to control the OSCs.

Because the accident is a result of several failures, it is difficult to hand overall 'responsibility' for the accident to one company. Various investigations point to flaws in the actions of a number of companies. Congressional investigators accused BP in relation to some of its key decisions, such as

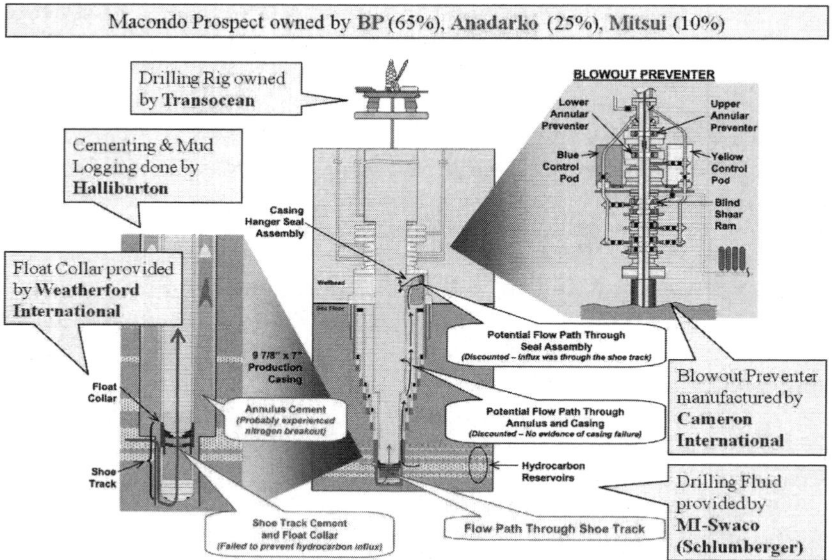

Figure 3.2 Companies Involved in the Macondo Well

Source: Author for descriptions of tasks performed by OSCs. BP's 'Deepwater Horizon: Accident Investigation Report' for the Macondo well diagram (BP 2010b).

choosing the riskier option in well design and in the number of centralizers[22] (Crooks and Fifield 2010). Another inquiry[23] by the National Commission set up by the president has also identified several central mistakes made by BP, including BP's misinterpretation of the vital 'negative pressure test', used to see whether the well had been properly sealed with cement before the rig moved away (National Commission 2011). In the same report Halliburton is criticized for its failure to ensure that the cement had been tested properly and Transocean for failing to inform its crew about the lessons learned from a 'near miss' in the North Sea four months prior to the Macondo accident. The report states that most of the mistakes and oversights can be traced back to the 'failure of management', in other words 'failure of service integration' (Crooks and Pfeifer 2011).

On the other hand, BP's internal inquiry reveals that BP's failures were primarily a failure to 'manage its contractors properly' and prevent the mistakes that they made. For example, the report states that the first critical factor in the accident was the failed cement job. This allowed oil and gas to leak into the well when it should have been sealed off. Halliburton comes under fire in the report due to its weaknesses in the cement design and testing, quality assurance and risk assessment. Transocean is also heavily criticized due to the misreading of the pressure tests. Despite surprising pressure readings, both BP and Transocean concluded incorrectly that the test had been a success and the well was secure. Transocean is also seen as responsible for the failure of the BOP because it had not tested the BOP before its installation and ignored previous maintenance problems (Crooks and Pfeifer 2010a). In addition, an internal audit by BP prior to the accident found 390 overdue maintenance problems on the Deepwater Horizon rig that Transocean was utilizing. However, despite the ongoing maintenance requirements, it was BP that ultimately decided the rig was fit for service (Pfeifer and McNulty 2010a).

It is impossible to apportion final accountability within the context of this book. However, we can deduce that the accident shows a problem of integration for the services in the supply chain. Andrew Hill shares this view, pointing out that the Deepwater disaster teaches lessons about how to run the ultimate 'extended enterprise'. He supports his view with a report published by the US National Commission in January 2011. The report stated that the poor integration of 'corporate cultures, internal procedures and decision making protocols' of the different companies working together on the Deepwater Horizon rig was one of the causes of the explosion. According to Hill, the more closely the producers, suppliers and customers collaborate, the smaller the gaps in communication and understanding of goals, values or the best execution of work will be, and it is these failures that can undermine the whole undertaking (Hill 2011).

The accident highlights issues of integration on two fronts. From an operational point of view, the accident provides evidence for the problems with effective coordination and integration of day-to-day activities from cementing to well drilling, and from BOP to well testing, which were among the causes of the accident. For example, BP's failure to spot that Transocean had missed the testing of the BOP shows that BP did not fully consider the risks created in each segment of the supply chain. Day-to-day integration and coordination problems were also highlighted by the Presidential Commission report, set up by Obama, which concluded that contractors failed to conduct proper risk assessment of the changes to the well and drilling procedure. They also failed to share information with BP and with each other. Furthermore, the report states that BP's fundamental mistake was its failure to exercise special caution and to direct its contractors to be especially vigilant (Crooks and Pfeifer 2011). BP as the system integrator and assembler of all services showed problems of system integration.

More profoundly, in the long term, the accident demonstrates what might happen when one segment of services—services related to safety—does not keep pace with the progress of other services in an industry. In the oil industry, while deepwater drilling services have significantly advanced, the safety industry has not followed suit. Technology has moved away from underlying safety systems. BOP, a supposedly fail-safe device which acts as the last line of defence against oil and gas spills, failed—showing that the entire industry was based on false security. Contingency plans were inadequate. Fundamental seabed equipment did not exist for undersea accidents, showing that safety technology has not progressed in line with the drilling technology.

The oil industry has been criticized for developing technologies to drill at great depths without adequately working out what would be required to stop a spill a mile under the sea. As the industry's capabilities have grown, companies have pushed into deeper waters and into more extreme conditions without fully understanding the dangers and developing additional safeguards. For example, a Halliburton presentation in 2009 warned of the risks involved in cementing deepwater wells. It listed the characteristics of a good cement job and noted that 'conditions in deepwater wells are not conducive to achieving all of these objectives simultaneously' (Crooks and McNulty 2010). A number of interviewees concluded that the fact that safety technology has not progressed alongside drilling technology reveals long-term gaps in the integration of services and technologies. It can be concluded that system integrators have not guided technological progress well enough to ensure that all services are up to speed with the required level of technical advancement.

Beside the causes, one of the most debated issues following the accident is the placement of liability. The question of liability complicates the relationship of IOCs, both with OSCs and with JV partners.

With regard to the JV partners, it is widely accepted that the operator of the field has the ultimate responsibility for dealing with the problem. As a general rule, junior partners pay for their share of investment and are responsible for their share of costs and liabilities associated with spills and accidents, unless an operator is found to have been grossly negligent (McNulty and Hoyos 2010). However in specific cases, where the final liability will reside—be it with the operator, contractors or JV partners—is seen as a matter of contractual relationship in each JV. Nevertheless, several interviewees mentioned that no company would take the role of operator if all risks remained with the operator of the field. In BP's accident, the US government inquiry has found no evidence that partners were involved in decisions. Decisions related to the design and drilling of the Macondo well were led by BP (McNulty and Crooks 2011). However, the joint venture contractual framework stipulates that all three partners were responsible parties. In line with the contracts, BP has reached a settlement agreement with both partners (Anadarko Petroleum for $4bn and Moex Offshore for $1.1bn) (Wembridge and Pfeifer 2011).

With regard to service providers, the determination of where liability lies (i.e., whether with the final decision maker and thus the integrator, or with the implementer hence the service providers) has extensive implications regarding company boundaries between the service providers and integrator.

To drill a well, the IOC (in this case BP) brings together a group of specialist companies of service and equipment providers—which on Deepwater Horizon included Transocean (rig owner and operator), Halliburton (cemented the well) and MI Swaco (provided drilling fluids). In the BP accident, the equipment providers assumed their portion of responsibility and settled with BP. Cameron, the manufacturer of BOP, agreed to a settlement of $250m and Weatherford International, the maker of the float collar, settled for $75m. MI Swaco, which was accused by BP of failing to monitor the mud and spacer solutions, also settled for an undisclosed sum.

Conversely, Transocean and Halliburton hold a different view and have been in dispute with BP regarding the indemnities. Transocean has been accused and held liable by BP ever since the accident first occurred. In his first public statements, Tony Hayward said that it was Transocean's people and processes at work on the rig and it was not a BP accident[24] (Crooks and Andrew 2010). BP stated that while the spill was its responsibility, it was not its accident because a subcontractor had been running the rig. Lamar Mckay, head of BP America, highlighted that 'Transocean had the responsibility for the safety of the equipment, including the blowout preventer, which was supposed to have cut off the well after the explosion. The systems are intended to fail-close and be fail-safe' (Lamar McKay quoted in Fifield 2010). However, this statement ignores BP's role as a supervisor and coordinator of the project. Since its initial reaction, BP has accepted that its staff was involved in key decisions on the rig (FT Weekend 2010). Steve Norman, CEO of Transocean, said his company was operating under BP's orders. He added that 'all offshore oil and gas productions begin and end with the operator' (Crooks and McNulty 2010). BP and Transocean are still in dispute regarding each other's role. The allocation of responsibilities as a result of legal proceedings will redefine the relationship between the system integrator and its suppliers. In other words, the courts will decide what the boundaries of the firm are.

With regard to Halliburton, a few reports on the disaster point to flaws in Halliburton's oil-well cement. Questions were asked about whether the unstable cement caused the accident, and if so, whose responsibility it was to check the quality and safety of the cementing job. The US National Commission report states that three out of the four cement tests conducted before the blowout showed the cement was unstable. Some of the test results were provided to BP, and others were not. These findings raise questions as to why Halliburton's employees did not act on the information (Pfeifer et al. 2010). According to the investigators, in March 2010 Halliburton provided data to BP taken from one of two tests, which showed that a foam slurry design very similar to the one actually pumped into the Macondo well would be unstable. Neither BP nor Halliburton acted upon this data. However, the data appeared in a technical report along with other information, and there is no indication that Halliburton highlighted the negative results to BP (Pfeifer et al. 2010). Inquiries have found that the failure of the cement seal to prevent leakage of oil and gas into the well as the rig prepared to move away was one of the principal causes of the accident. BP and Halliburton disagree over how much responsibility they each bear for that failure (Crooks 2011a).

BP has sued Halliburton and reasserted its conviction that the OSC should be liable for the costs of the accident. Tim Probert, Halliburton's Health, Safety and Environmental Chief, said, 'Halliburton is confident that the cementing work on the well was completed in accordance with the requirements of the well owner's well construction plan' (Tim Probert quoted in Fifield 2010). Halliburton refused a settlement, insisting it had followed BP's instructions. The case is still before the law courts. Doug Kysar, a professor at Yale Law School, said that—even if other companies should take some blame—a court might still find BP liable to pay damages 'because of the control it had over its contractors'. It would then be forced to pursue the associated companies for compensation (Crooks and Pfeifer 2010a).

The US government has brought charges against the three companies at the heart of the accident, BP, Transocean and Halliburton, for breaches of offshore regulations. The citation of additional companies is welcomed by BP, which has argued that the failures contributing to the accident are the responsibility of several companies and are not limited to BP (Crooks and McNulty 2011a).

A settlement or liability following the legal process could upend the business model in which operators bear most of the risks (Financial Times 2012). In the majority of cases, contracts between IOCs and OSCs stipulate that service providers do not participate in the indemnity unless their gross negligence can be proven. For example, in the service contract between BP and Transocean, BP agreed to 'protect, defend and indemnify and hold harmless' Transocean from 'all claims, demands, causes of action, damages, costs and expenses'.[25] According to Crooks (2011c), the contract specified that the indemnity held even in the event of gross negligence. However BP has defended that Transocean materially breached the contract and therefore cannot take advantage of indemnification clauses. Transocean has stated that if it is forced to bear a high proportion of the costs in spite of its contract with BP, it would make operations in the oil industry impossible because service companies would be unwilling to bear the risks of an accident. Transocean said that 'the responsibility for hydrocarbons discharged from a well lies solely with its owner and operators [in this case BP]' and that it had legal protection reinforced in its contracts (Crooks and Pfeifer 2010b).

The sharing of liability is becoming a major issue in contract negotiations. According to certain OSCs, there is a tendency for IOCs to push liabilities onto OSCs during contractual negotiations. OSCs strongly oppose this tendency, underlining that contractors are not decision makers and that liabilities should remain with the IOCs as they are the project managers. Certain IOCs agree with this, recognizing that if Halliburton and Transocean are held accountable, the relationship between IOCs and OSCs will become highly complicated. On the other hand, if the majority of liability continues to lie with the IOCs, it is expected that the system integrator will further increase its influence on the operations of the service providers in order to better control risks, hence increasingly blurring company borders.

Following the accident, Tony Hayward argued that the industry can cut the risk of serious accidents but this may mean changing the way the drilling industry operates; he further reasoned that the model of bringing together several specialist companies may have to change. He said, 'This is not about BP and Transocean. Transocean is a very good drilling contractor but we have to ask how much further we can drive the risk down'. He also stated that the accident raises the question of whether the existing industrial paradigm is right for the future. Writing an opinion piece in the *Wall Street Journal*, Hayward said, 'The industry has for decades relied on outsourcing work to specialized contractors. But the question after the Macondo accident is how all involved parties can work even more closely together to better understand and significantly reduce the various risks associated with drilling operations' (Hayward 2010b). He also put forward the possibility that in the future BP would operate its own rigs working in deepwater. Shell has already moved in a similar direction by creating a JV with Frontier Drilling (Crooks and Luce 2010). They have jointly developed a new drillship 'bully rig', to be used in deepwater and arctic conditions, which is easier to manoeuvre and more energy efficient than a traditional drillship (Shell 2011b). Shell executives believe that by building some of its own rigs through the joint venture, it can reduce its dependence on subcontractors and retain technological control. Following the Macondo accident, if BP decides to move further in the same direction, it will mean a radical change for the existing industrial structure (Crooks and Luce 2010).

According to one interviewee, the company managing the project, integrating services and taking responsibility and liability should be the IOC. However, the interviewee went on to say that OSCs should recommend certain practices and warn IOCs if safety practices are breached. For example, an OSC has advised its employees to recommend services and safety practices to IOCs and to inform their top management if their recommendation is not followed. However, it is not very clear if this particular OSC's course of action would have avoided the Macondo accident because a number of discussions had already taken place between BP and related OSCs prior to the accident. According to the *Financial Times*, there was a dispute between a top BP employee and a top Transocean official on the day of the blast about a decision to remove heavy drilling mud from the drill pipe and replace it with water. BP has said the decision to push ahead with the procedure, despite the questionable pressure reading, was a fundamental mistake (Kirshgaessner 2010). Similarly, Halliburton said that 'it had raised concerns about the number of centralisers to be run and made recommendations regarding the cementing services provided; however ultimately Halliburton acted on the decisions of and at the explicit direction of the well owner'(Pfeifer and McNulty 2010b). BP opposed Halliburton's view and said in a statement, 'Halliburton was fully aware of the decision to use six centralisers. If Halliburton harboured any significant concerns about the safety of the operation, then it had a moral and legal responsibility to refuse to perform the job' (Pfeifer and McNulty 2010b). Therefore, it is questionable if recommending best practices and disputing concerns would have avoided the accident.

Closely related to matters of liability, the accident raises questions over who should bear the catastrophic event risk. In general, OSCs are seen as too small to take catastrophic risks that have a low probability but high impact. For instance, Transocean is the world's largest drilling company but the BP accident is still too big in comparison to the size of Transocean's balance sheet. Even if contractually risks could be passed to an OSC, in reality risks would remain with the IOCs due to their size. These types of risk are considered beyond the capacity of OSCs. In addition, when OSCs take some of the risks, they try to pass these on to their subcontractors on the supply chain. Risk is not managed well when it falls on the last company at the end of the supply chain. Even if these smaller companies assume responsibility contractually, IOCs remain ultimately responsible due to their size and their ability to pay. Liability rests with the operator in the event that a company on the supply chain defaults. Thus, IOCs need to control the supply chain to a greater degree for the efficient management of risk.

Because prevention is seen as the best management of risk, operational and contractual risk will be revised by IOCs following the Macondo accident. More controls will be implemented to avoid Macondo-type accidents, and better accident response mechanisms will be defined in advance. An example of this already in practice can be seen in the recent contracts between IOCs and OSCs, which include provisions related to the cost of unused drilling capacity in case of a moratorium, similar to the six-month moratorium declared in the Gulf of Mexico following the Macondo accident.

With regard to the consequences of the accident, it is widely accepted that the accident will impact industry structure and increase the need for big service companies. The additional burden on companies due to more stringent standards and regulations is likely to favour large service companies. However it is worth noting that the two main service companies involved in the accident are the largest players in their fields.

According to an interviewee, IOCs have realized that they are very vulnerable to 'the mistakes' of OSCs. BP's internal inquiry shifts much of the burden of responsibility for the accident onto its contractors, notably Transocean and Halliburton. According to a few interviewees, in response, IOCs need to be more involved in day-to-day management of OSC activities and use more third-party verification.

Following the accident, host countries have imposed tougher rules and tighter controls on the drilling activities of IOCs. Countries like Brazil have imposed new rules to prevent spills (Person 2011). In order to better manage risks, IOCs pass these tougher rules along to OSCs and go even further by applying stricter criteria for awarding drilling contracts that require stringent equipment specifications and drilling procedures, putting increased emphasis on the competence of the personnel in the field. IOCs also set up voluntary standards on their side. For example, BP has implemented a set of voluntary drilling standards in the Gulf of Mexico that go beyond the existing regulatory requirements, including third-party verification of testing and maintenance of BOPs (Pfeifer 2011). Requirements such as checks on BOPs indicate that preparations for a worst-case scenario are much more thorough than they were before the disaster (Crooks and McNulty 2011b).

These procedures and requirements have direct impacts on OSCs. For example, the financial results of Transocean reveal that five of its deepwater rigs underperformed in Q32011 mainly because they needed repairs or maintenance to their BOPs (Pfeifer 2012). These rules are not limited to the services companies but also extend to equipment manufacturers. Even if IOCs do not tender the required equipment piece by piece, they would exercise more control over each piece and are likely apply a 'preferred supplier' approach with greater standardization for each item. Central decision making and procurement could become even stronger in order to reduce the variation of risks across contracts and to better control these risks. The aim will be to minimize risks everywhere and know exactly how much risk the company is taking on.

The accident will bring more centralization to outsourcing, which will be accompanied by greater control and a smaller number of suppliers. In addition, it will increase the cost of operations and insurance fees due to additional regulations and standards, and is likely to cause difficulties for smaller companies. Furthermore, the large scale of liabilities will increase the need for big companies with a good safety track record in the field of deep sea development. Robin West, chairman of PFC Energy, believes that over time companies with stronger safety records and stronger balance sheets will differentiate themselves (McNulty and Blas 2010). Therefore, further consolidation of suppliers is expected. The intense supervision, the greater scrutiny and the premium on the quality of services may well reduce the number of suppliers and help larger service companies (Hoyos and Thomas 2010). The accident may prompt IOCs to focus on using the latest and safest drilling ships and rigs, boosting sales of such vessels. IOCs will be more demanding regarding the aptitude of OSCs and their choice of operators. Only a few companies will be able to meet the standards set by the oil industry and host countries. Companies like Sea Drill, which has one of the market's most up-to-date fleets, are expected to benefit from the new requirements (Wright 2010). Partners will exercise rigorous control over the operator and the operator itself will require partners to contribute to the payment of damages in case of accident.

The accident will also impact smaller oil companies. IOCs are large enough and sufficiently diversified to cope with a moratorium. However, leaving a deepwater rig, at a cost of approximately $500,000 per day, idle over an extended period of time would cause severe financial problems for small oil companies such as Murphy Noble Energy and Kerr McGee, that may not have other global projects they can move rigs to, and this prospect could force such companies out of deepwater projects (McNulty and Blas 2010). In addition, the fact that BP has been forced to set up a $20 billion[26] fund has effectively removed the previous $75 million legal cap on liabilities for economic damage caused by an oil spill (Economist 2010). Considering the 'endless' liability, very few companies will be able to operate in the US (Hoyos and Thomas 2010).

As outlined above, the accident will have considerable impact on the oil industry, which plans to invest almost $170 billion from 2010 to 2014 globally in deep and ultra-deepwater developments, according to Douglas Westwood consultancy. Most of the spending will focus on the 'golden triangle' in deep waters of the Gulf of Mexico, off Angola and Nigeria in West Africa and off Brazil. Deep offshore exploration is not an optional activity for IOCs; it is a core capability. In the future, they will have to increasingly explore in deep waters. The BP accident shows the difficulty of rectifying problems that occur at great depths. Therefore, countries need to consider the best mechanisms to deal with rare but catastrophic accidents (Financial Times 2010a). Increased regulation and tighter controls will have a global impact (McNulty and Blas 2010).

The impact on the overall industry has initiated some heated debates among IOCs. The CEO of Exxon has accused BP of doing a great disservice to the oil industry by suggesting that the disaster is not a 'black swan' event and that it has implications for all IOCs. According to Exxon, the cause of the accident was a breakdown in management supervision, which is not an industry-wide problem (Crooks 2011b). Other IOCs have argued that the accident was preventable and their own safety systems are robust enough and do not need any significant reform. For example, Pete Slaiby, the VP of Shell Alaska, stated that 'the Gulf of Mexico may have been a wakeup call for some but not for Shell'. Similarly, the CEO of Exxon said, 'It appears clear to me that a number of design standards that I would consider to be the industry norm were not followed. . . . We would not have drilled the well the way they did' (Pete Slaiby quoted in Crooks et al. 2010).

Despite IOCs debating the 'mistakes specific to BP' and the impact of the accident on general industry practices, effects have already started to be felt across the industry with enhanced safety requirements and tighter control of OSCs' activities by IOCs. One area where companies have already begun to make progress is in developing spill-response systems, which were inadequate at the time of the accident. Tony Hayward admitted that BP did not have all the tools in its toolkit to stop a blowout on the seabed under 5,000 feet of water. Following the accident four major oil companies (Exxon Mobil, Chevron, Conoco Philips and Shell) committed $1 billion to fund the launch of a nonprofit JV, Marine Well Containment Company. The aim is to develop a spill response system that can be mobilized within twenty-four hours of an incident, and be used in deepwater depths of up to

10,000 feet, with a capacity to contain up to 100,000 barrels of oil per day (Coombs 2010).

With regard to BP, several court cases are still ongoing at the time of writing. The greatest unknown remains the outcome of the Department of Justice's investigation into the spill and whether BP will be found guilty of gross negligence. BP has already set aside $39.9 billion to cover the cost of the spill. In case of gross negligence, BP could face an additional penalty of $21 billion or more (Crooks 2010b). The final impact of the accident on BP will therefore be decided in the courts.

While the final court decision is still pending, analysis of share prices following the accident shows that financial markets consider BP and Transocean to be the primary culprits. The share price of all the companies went down after the accident. Halliburton has recovered since the accident as of April 20, 2012. However, BP and Transocean share prices are trading at heavy losses as of May 2012.

Despite the ongoing court cases, BP has been cleared by US regulators to return to the Gulf, having received the first permit to drill a new well on the Kaskida prospect. In order to be awarded the drilling permit, BP has met all safety requirements and has adhered to voluntary standards that go beyond the new regulatory requirements (Pfeifer 2011). Bob Dudley, the new CEO of BP, said that the strategy of BP following the accident is to prioritize safety and risk management and thus become comfortable managing risks while retaining the spirit of an entrepreneurial operator (Pfeifer 2010).

In conclusion, the BP accident shows how system integrators need to manage the activities of suppliers in great detail on a day-to-day basis and

Table 3.3 Stock Indices of OSCs Involved in the Macondo Accident and Selected IOCs

Company	20/04/2010	20/04/2011	20/04/2012
BP	100	71	66
Shell	100	111	114
Total*	100	103	82
Schlumberger	100	130	106
Halliburton	100	149	100
Transocean	100	82	55
Weatherford International	100	123	80
Cameron International	100	119	106

Source: Compiled by the author using Bloomberg data (as of May 23, 2012). Used with permission of Bloomberg L.P. Copyright© 2015. All rights reserved.

Note: Share prices are indexed to 100 on April 20, 2010, the day of the BP accident.

* *Total's share price decline in 2012 is mainly due to a gas leak that occurred in the Total-operated Elgin Franklin gas field in March 2012.*

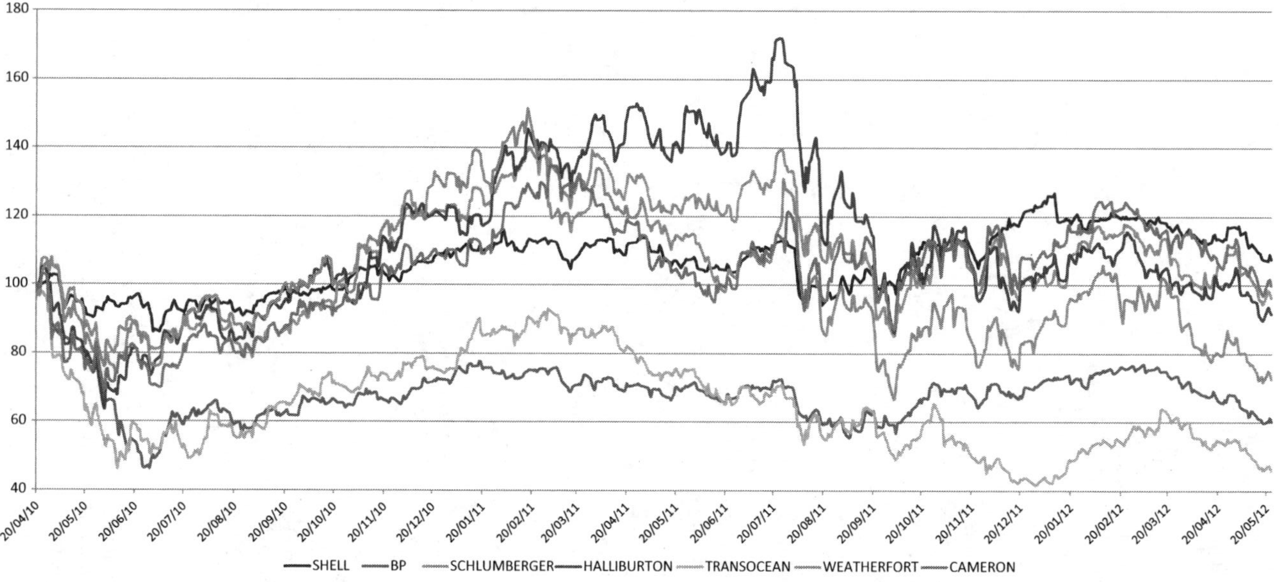

Figure 3.3 Stock Indices of OSCs Involved in the Macondo Accident and Selected IOCs

Source: Compiled by the author using Bloomberg data (as of May 23, 2012). Used with permission of Bloomberg L.P. Copyright© 2015. All rights reserved.

how oversights can have massive repercussions. The incident has highlighted many problems associated with system integration, definition of boundaries and the extent of liability in case of accidents. In essence, all specifications and decisions regarding the well are determined by the system integrator. As Halliburton clearly stated, the work it performed on the Macondo well was completed in accordance with the well owner's specifications (Pfeifer et al. 2010). A profound level of system integration and the extent of liabilities are clearly shown by the BP accident; irreversible problems may occur when outsourced activities are not integrated effectively. A system integrator needs to give its full attention to a service or a function provided by its suppliers. Despite the supplier's providing the services, the system integrator retains responsibility and accountability (Stern 2010). Following the accident, most IOCs confirmed plans to review their use of third-party contractors and the control they exercise over them. As the accident proved, contractors operate in the name of their customer, and companies are fully responsible for what their contractors do (Skapinker 2010). In the BP accident, court decisions will show where the boundary of the firm is. If BP is held responsible for all liabilities by legal courts, the courts will be confirming that BP incorporates all other companies it contracts to into one extended company. The Macondo accident demonstrates how deeply integrated companies on an oil rig are and how problems of integration may have disastrous consequences.

NOTES

1. For example, they decide whether an FPSO or a jack-up will be used.
2. 'The IPB is a flexible production riser including thermal insulation layers, additional hoses for gas lift or other services, active heating through electric cables and fluid temperature monitoring with optical fibres. The IPB allows for high-level flow assurance of hydrocarbon fluids in difficult conditions (viscous oil, deepwater, pressure constraints, etc.) from well heads to surface treatment units' (Technip 2012).
3. Wide Azimuth is a seismic method that enables oil companies to gain better illumination and resolution in seismic data by ensuring a wide-azimuth recording and dense sampling to obtain a better representation of the seismic wave field (CGGVeritas 2012).
4. The Patent Board is patent ratings partner for the *Wall Street Journal*, with a weekly Patent Scorecard column for the WSJ Market Data. 'The Patent Board Top 20 ranks corporate innovation using a series of metrics to determine patent quality, technological strength and breadth of impact' (Patent Board 2012). '*Patents Granted*—equals the number of US patents granted in a given year, excluding design and other special-case inventions. *Science Strength*—ranking measure to indicate how much a company uses science in building its patent portfolio with a combined measure of science and quantity. *Innovation Cycle Time*—indicates whether a patent or patent portfolio is building off newer or older inventions. *Industry Impact*—indicates the extent to which others are building upon a portfolio of issued US utility patents as compared to the total set of utility patents. *Technology Strength*—ranking measure to indicate an overall strength of the company's patent portfolio holdings with a combined

measure of quality and quantity. *Research Intensity*—indicates the extent to which a portfolio includes patents with above average Science Linkage as compared to the control group.' The Patent Scorecard includes all US utility patents that were represented by each entity (IPIQ 2012).

5. Shell was the top spending company in R&D among IOCs between 2007 and 2011 on a cumulative basis (Shell 2011b).
6. That compares with typical proportions of 15% for technology and pharmaceutical companies, and 4 to 5% for motor companies (Crooks 2008).
7. IHS CERA Upstream Capital Cost Index (UCCI) indicates the cost associated with construction new oil and gas facilities. The values are indexed to the year 2000, meaning that a piece of equipment that cost $100 in 2000 would cost $230 in 2008. Please see Figure 3.1 on IHS CERA UCCI for further information.
8. Expandable tubulars is an emerging tubular technology. In-situ expansion of this type of tubular material allows for preservation of borehole size while reducing pipe tapering, which can be especially useful in deep wells. Expandable tubulars can also be used to repair wells with damaged completions. Expandable tubulars was a concept first developed by Royal Dutch Shell around 1993. Commercial application of this technology occurred in November 1999 as a solid open hole expandable liner in the Gulf of Mexico (Cassidy and Butterfield 2002).
9. Surface Slice interpretation allows interpreters to scan the 3D shape of the dome through horizontal slides that resemble a series of contour maps. It is a new fast-volume interpretation tool.
10. Saudi Aramco history: 1933—Californian Arabian Standard Oil 100% Chevron; 1936—Chevron 50%, Texaco 50%; 1938—oil found; 1944—renamed Arabian American Oil (Aramco), Chevron 30%, Texaco 30%, Exxon 30%, Mobil 10%; 1974—nationalization: state 60%, Chevron 12%, Texaco 12%, Exxon 12%, Mobil 4%; 1980—state 100% (Hermann et al. 2010).
11. Sources for share in access to oil reserves and in oil production are indicated in section 2.1.2. Hard data for share in new oil field development is not available. The figure 30% was stated by a high-ranking interviewee.
12. The US is a notable exception where individuals may own mineral rights.
13. In effect, a buy-back transaction is a method of trade where plants, machinery, production equipment and technology are supplied (by a domestic or foreign private firm), in exchange for the goods that will be produced directly or indirectly by means of such facilities. Investors repatriate the return on the investment (at a pre-agreed fixed rate) through goods and services produced by the project. The buy-back scheme is a formula used by the Iranian government to attract foreign investment. Because Iran's revolutionary ideology and constitution forbid granting 'concessions', buy-back contracts were devised as a compromise solution in 1989.
14. For example, BP's and Shell's share prices increased from 169p and €8.42 in January 1990 to 596p and €29.73 by the beginning of 2000. Both are lower in 2015 with BP hovering around 411p and Shell around €27.
15. For example, the share prices of Schlumberger and Halliburton increased from $11.02 and $10.22 on January 2, 1990, to $27.47 and $19.16 on January 3, 2000. Figures stood at $85.41 and $39.33 on January 1, 2015.
16. In a reservoir below the sea, the earth is under enormous pressure. Puncturing the earth at these depths is like sticking a pin in a balloon full of water, and liquid tends to shoot out. A blowout is an uncontrolled release of oil or natural gas when the pressure is not controlled. Gasses at depth escape the well and ignite. Prior to the advance of pressure-controlling equipment in 1920, blowouts were common, known as the famous 'oil gusher'. Over time, the

oil industry has been successful in managing pressure problems. This requires pressure drillers pumping 'mud' (specialized fluid) down the pipes carrying the drill into the well. Balancing the opposing pressure is a highly skilled job assisted in modern rigs by sensors reporting on the conditions of the well.

17. Workers were preparing to detach the Deepwater Horizon rig from the well it had drilled. A steel casing had been inserted into the hole and cemented by Halliburton. The final stage would have been to put in cement to plug the well so that it could be left safely while BP decided how to start production.
18. One of the first pieces of safety equipment designed for oil and gas production, the BOP is a large valve at the top of a well that can be closed if the drilling crew loses control of oil or natural gas while drilling or performing a work-over. If these fluids manage to enter the wellbore, they may threaten the safety of the rig and crew. BOPs are described as the main barriers protecting human life, capital equipment and the environment by Melvyn Whitby of Cameron International, a leading maker of BOPs.
19. The Exxon Valdez accident took place on March 24, 1989, when an Exxon super tanker ran aground in Alaska. This was the largest oil spill in US waters before the BP accident in 2010.
20. A fire and explosion occurred in a Texas refinery operated by BP on March 23, 2005.
21. It involved a well integrity failure, followed by a loss of hydrostatic control of the well. This was followed by a failure to control the flow from the well with the BOP equipment, allowing the release and subsequent ignition of hydrocarbons. Ultimately BOP emergency functions failed to seal the well after the initial explosions. BP's inquiry identifies eight critical factors including substandard cement work, misinterpretation of the results of a key pressure test and faulty valves that failed to stop the oil and gas escaping.
22. Congressional investigators have accused BP relating to five decisions: 1) BP chose the riskier option when it was installing the casing or lining in the well the day before the accident. 2) BP failed to use enough centralizers to keep the casing in the centre of the borehole as it was lowered into the well. 3) BP and its contractors failed to run an acoustic test to check whether the cement used to attach the casing to the rock walls of the well had formed a seal to prevent the gas leak. 4) BP did not properly pump enough drilling fluid through the well to check for and remove pockets of gas before cementing the well. 5) BP failed to secure the top of the well properly with a lock-down sleeve to keep it in place and seal it tightly, thereby allowing oil and gas to leak out and rise up the pipe to the rig at the surface (US House Energy and Commerce Subcommittee on Oversight and Investigation, quoted in Crooks and Fifield 2010).
23. The report also highlighted BP's decision to replace the heavy drilling 'mud' in the pipes leading to the rig with lighter seawater before a seal had been placed in the top of the well.
24. He told the BBC: 'This was not our drilling rig. It was not our equipment; it was not our people, our systems and our processes. This was Transocean's rig, their systems, their people, their equipment.'
25. Most drilling contracts share liabilities under 'mutual hold harmless' in other words 'knock for knock' indemnities regime. Please see section 4.2.3 for further details.
26. The likely cost of the accident to BP will come in three parts. The first element is the direct costs of plugging the well and cleaning up the pollution. Then there are penalties under the Clean Water Act. Finally, there is compensation for lost economic activity, lost federal, state and local taxes and damage to the environment (Economist 2010).

REFERENCES

Baxter, K. 2009. "Technip & Schlumberger Agree Subsea Pipes JV." *Arabian Oil & Gas* [Online], 23 November. Accessed 7 September 2012. http://www.arabianoilandgas.com/article-6533-technip-schlumberger-agree-subsea-pipes-jv/.

BBC. 2009. "Iraq Oil Development Rights Contracts Awarded." *BBC* [Online], 11 December. Accessed 1 September 2011. http://news.bbc.co.uk/1/hi/8407274.stm.

Bourque, J., F. Tuedor, L. Turner, S. Gomersall, P. Hughes, R. Klein, G. Nilsen, and D. Taylor. 1997. "Business Solutions for E&P through Integrated Project Management." *Oilfield Review Schlumberger* 9: 34–49.

BP. 2010a. "Deepwater Horizon Containment and Response: Harnessing Capabilities and Lessons Learned." 1 September.

BP. 2010b. "Deepwater Horizon: Accident Investigation Report." *BP* [Online], 8 September 2010. Accessed 22 May 2015. http://www.bp.com/content/dam/bp/pdf/gulf-of-mexico/Deepwater_Horizon_Accident_Investigation_Report.pdf.

BPI. 2012. "About the BP Institute." BP Institute for Multiphase Flow, University of Cambridge. Accessed 1 September 2012. http://www.bpi.cam.ac.uk/intro/.

Cassidy, J., and C. Butterfield. 2002. "Electrochemical Investigation of Oilfield Fluid Corrosion on Expanded Casing." NACE International.

CGGVeritas. 2012. "What Is Wide Azimuth (WAZ)?" Accessed 2 March 2012. http://www.cggveritas.com/default.aspx?cid=1833&lang=1.

Coombs, B. 2010. "Four Energy Companies Form Oil Spill Reaction Team." *CNBC* [Online], 21 July. Accessed 30 December 2012. http://www.cnbc.com/id/38352657/Four_Energy_Companies_Form_Oil_Spill_Reaction_Team.

Crooks, E. 2008. "Shell Sets Pace as Big Oil Lifts R&D Spend." *The Financial Times*, 29 July.

Crooks, E. 2010a. "BP Leak Just a Bump in Road for further Oil Industry Exploration." *The Financial Times*, 19 September.

Crooks, E. 2010b. "Cost of Deepwater Horizon Rests with the US Courts." *The Financial Times*, 17 December.

Crooks, E. 2011a. "BP Blames Halliburton Destroyed Deepwater Horizon Evidence." *The Financial Times*, 6 December.

Crooks, E. 2011b. "Exxon's Tillerson Rejects BP Take on Oil Spill." *The Financial Times*, 10 March.

Crooks, E. 2011c. "Transocean Lines Up BP Claim." *The Financial Times*, 31 October.

Crooks, E., and E.-J. Andrew. 2010. "BP Counts High Cost of Clean-up and Blow to Image." *The Financial Times*, 5 May.

Crooks, E., and A. Fifield. 2010. "Hayward Responses Raise Hackles." *The Financial Times*, 18 June.

Crooks, E., and E. Luce. 2010. "Industry Can Cut Accident Risks, Says BP." *The Financial Times*, 3 June.

Crooks, E., and S. McNulty. 2010. "A Spreading Stain." *The Financial Times*, 7 May.

Crooks, E., and S. McNulty. 2011a. "BP, Halliburton and Transocean Charged over Deepwater Disaster." *The Financial Times*, 13 October.

Crooks, E., and S. McNulty. 2011b. "Lease Sale Offers Revival Hope for Gulf Production." *The Financial Times*, 14 November.

Crooks, E., and S. Pfeifer. 2010a. "BP Shares Out Thunder for Its Perfect Storm." *The Financial Times*, 9 September.

Crooks, E., and S. Pfeifer. 2010b. "BP Sued in $21 Billion US Gulf Spill Action." *The Financial Times*, 16 December.

Crooks, E., and S. Pfeifer. 2011. "BP Investors Shrugs Off Critics." *The Financial Times*, 7 January.

Crooks, E., C. Hoyos, and S. McNulty. 2010. "Tougher US Rules on Drilling Loom Large." *The Financial Times*, 22 July.

Daly, J.C.K. 2012. "Mexico to Privatize State Oil Company Pemex?" *Rigzone* [Online], 14 December. Accessed 20 December 2012. http://www.rigzone.com/news/article.asp?hpf=1&a_id=122810.

Dirksen, R. 2009. "Hostile Drilling Environments Require New Approach." EPMAG, Hart Energy, 1 August. Accessed 15 December 2011. http://www.epmag.com/Production-Drilling/Hostile-drilling-environments-require-approach_42842.

Economist. 2010. "The Oil Well and the Damage Done." *The Economist*, 17 June.

ExxonMobil. 2011. "Summary 2011 Annual Report." Accessed 5 September 2012. http://www.exxonmobil.com/Corporate/Files/news_pub_sar2011.pdf.

Farnejad, H. 2006. "How Competitive Is the Iranian Buy-back Contracts in Comparison to Contractual Production Sharing Fiscal Systems?" *CEPMLP Annual Review, University of Dundee* [Online]. Accessed 5 September 2012. http://www.dundee.ac.uk/cepmlp/car/html/CAR10_ARTICLE16B.PDF.

Fifield, A. 2010. "Senators Scorn Efforts to Pass the Blame." *The Financial Times*, 12 May.

Financial Times. 2010a. "Cleaning Up After the BP Spillage." *The Financial Times*, 27 May.

Financial Times. 2010b. "Oil Disaster Lessons." *The Financial Times*, 3 May.

Financial Times. 2012. "Lex Column." *The Financial Times*, 4 January.

Flaharty, G.R. 1999. "Shell and Baker Hughes Establish Joint Venture for Expanded-Tube Well Construction and Remediation Technology." *Oil and Gas Online* [Online], 8 February. Accessed 15 December 2011. http://www.oilandgasonline.com/doc.mvc/Shell-and-Baker-Hughes-Establish-Joint-Ventur-0001.

FT Weekend. 2010. "BP: The Inside Story." *The Financial Times*, 2 July.

Harman, B. 2007. "Oil and Gas Industry Primer." *Investopedia* [Online], 21 May. Accessed 25 May 2012. http://www.investopedia.com/articles/07/oil_gas.asp.

Hayward, T. 2010a. "RE." Speech at Cambridge Union Society, 10 November.

Hayward, T. 2010b. "What BP Is Doing About the Gulf Gusher?" *The Wall Street Journal*, 4 June.

Hermann, L., J. Copus, and J. Hubbard. 2008. *A Guide to Oil & Gas Industry*. Global Markets Research. London: Deutsche Bank.

Hermann, L., E. Dunphy, and J. Copus. 2010. *Oil & Gas for Beginners: A Guide to the Oil Industry*. Global Markets Research. London: Deutsche Bank.

Hill, A. 2011. "BP's Woes Are a Guide to Modern Executives." *The Financial Times*, 18 January.

Hoyos, C. 2010. "The Rigs Blow-out Preventer Holds the Key to What Went Wrong and Why." *The Financial Times*, 7 May.

Hoyos, C., and H. Thomas. 2010. "BP Fallout Threatens Smaller Operators." *The Financial Times*, 23 June.

IHS. 2015. "IHS CERA Upstream Capital Costs Index (UCCI)." Accessed 27 April 2015. http://www.ihs.com/info/cera/ihsindexes/index.aspx.

IPIQ. 2012. "Shell & Schlumberger Trade Spots at the Top." *IPIQ Global* [Online], 29 June. Accessed 5 July 2012. http://ipiqglobal.net/ing/sec_5/art.php?id=51. www.IPIQGlobal.com.

Kirshgaessner, S. 2010. "Mystery over Crew's Reaction Times." *The Financial Times*, 28 May.

McNulty, S., and C. Hoyos. 2010. "Oil Majors Review Ties with BP After the Spill." *The Financial Times*, 25 June.

McNulty, S., and E. Crooks. 2011. "Deepwater Continues to Reverberate." *The Financial Times*, 19 October.

McNulty, S., and J. Blas. 2010. "Big Oil Group Break Ranks with BP." *The Financial Times*, 4 June.

Menas. 2012. "Iran: Energy Industry Overview." Accessed 15 December 2012. http://www.menas.co.uk/localcontent/home.aspx?country=75&tab=industry.

National Commission. 2011. "Report to the President: Deepwater, the Gulf Oil Disaster and the Future of Offshore Drilling." National Commission on the BP Deepwater Horizon Oil Spill and Offshore Drilling, January.

Osmundsen, P., T. Sørenes, and A. Toft. 2009. "Oil Service Contracts, New Incentive Schemes to Promote Drilling Efficiency." Department of Industrial Economics and Risk Management, University of Stavanger. Accessed 5 September 2012. http://www1.uis.no/ansatt/odegaard/uis_wps_econ_fin/uis_wps_2009_7_osmundsen_sorenes_toft.pdf.

Patent Board. 2012. "The Top 20 Ranking, Energy & Environmental." Accessed 2 September 2012. http://patentboard.com/Home/tabid/38/Default.aspx.

Pfeifer, S. 2010. "BP Sees Rays of Hope After the Darkest Hour." *The Financial Times*, 3 November.

Pfeifer, S. 2011. "Regulators in US Clear BP to Return to the Gulf." *The Financial Times*, 27 October.

Pfeifer, S. 2012. "Transocean Books $1 Billion Macondo Charge." *The Financial Times*, 27 February.

Pfeifer, S., and S. McNulty. 2010a. "BP Listed 390 Problems on the Rig." *The Financial Times*, 24 August.

Pfeifer, S., and S. McNulty. 2010b. "E-mail from BP Rig Says Cement Job 'Went Well'." *The Financial Times*, 24 August.

Pfeifer, S., S. McNulty, and S. Kirchgaessner. 2010. "Halliburton and BP Knew the Risks Before Spill." *The Financial Times*, 29 October.

Reuters. 2011. "Técnicas Reunidas SA, Technip, JGC Corporation and Petrovietnam Construction JSC Obtain USD 5,000 Million Contract to Build Oil Refinery in Vietnam." *Reuters* [Online], 5 January. Accessed 5 September 2012. http://in.reuters.com/finance/stocks/TECF.PA/key-developments/article/2057835.

Saudi Aramco. 2010. "Saudi Aramco Ranked Top Oil Company." *Latest News* [Online]. Accessed 16 May 2012. http://www.saudiaramco.com/en/home/news/latest-news/2010/saudi-aramco-ranked-top-oil-company.html#news%257C%252Fen%252Fhome%252Fnews%252Flatest-news%252F2010%252Fsaudi-aramco-ranked-top-oil-company.baseajax.html.

Schlumberger. 2011. "Annual Report 2011." Accessed 26 September 2012. http://investorcenter.slb.com/phoenix.zhtml?c=97513&p=irol-reportsannual.

Schlumberger. 2012a. "Schlumberger Supply Chain Services." *Resources* [Online]. Accessed 16 May 2012. http://www.slb.com/resources/supply.aspx.

Schlumberger. 2012b. "Total Joins Chevron and Schlumberger Collaboration on Development of the INTERSECT Next-Generation Reservoir Simulator." *Schlumberger Press Release* [Online], 26 July. Accessed 16 August 2012. http://www.slb.com/news/press_releases/2012/2012_0726_intersect_collaboration_pr.aspx.

Schlumberger. 2012c. "Schlumberger Gould Research Center." *Research* [Online]. Accessed 22 May 2015. http://www.slb.com/about/rd/research/sgr.aspx.

Shell. 2011a. "Annual Report 2011." Royal Dutch Shell Plc. Accessed 26 September 2012. http://reports.shell.com/annual-report/2011/servicepages/welcome.php.

Shell. 2011b. "Briefing Note: 'Technology in the Arctic.' " April. Accessed 6 June 2012. http://www-static.shell.com/static/innovation/downloads/arctic/technology_in_the_arctic.pdf.

Shell. 2011c. "Investors Handbook 2007–2011: R&D Expenditure." Accessed 1 December 2012. http://reports.shell.com/investors-handbook/2011/projectstechnology/rdexpenditure.html.

Skapinker, M. 2010. "Memo to Board: We Need to Talk about BP." *The Financial Times*, 2 November.

Stern, S. 2010. "Outsource in Haste, Repent at Leisure." *The Financial Times*, 8 June.

Team, T. 2012a. "Shell's Record Transocean Deal Shows the Importance of Ultra-Deepwater." *Forbes* [Online], 10 February. Accessed 2 December 2012. http://www.forbes.com/sites/greatspeculations/2012/10/02/shells-record-transocean-deal-shows-the-importance-of-ultra-deepwater/.

Technip. 2012. "Flexible Pipe." Accessed 3 June 2012. http://www.technip.com/en/our-business/subsea/flexible-pipe.

Thurber, M., and P. Nolan. 2010. "On the State's Choice of Oil Company: Risk Management and the Frontier of the Petroleum Industry." Program on Energy and Sustainable Development, Stanford University: 99. Accessed 15 May 2012. http://iis-db.stanford.edu/pubs/23057/WP_99,_Nolan_Thurber,_Risk_and_the_Oil_Industry,_10_December_2010.pdf.

Thuriaux-Alemán, B., S. Salisbury, P.R. Dutto, and A.D. Little. 2010. "R&D Investment Trends and the Rise of NOCs." *Journal of Petroleum Technology, Society of Petroleum Engineers* 62: 30–32.

Total. 2003. "Girassol: A Stepping Stone for the Industry." Accessed 5 May 2008. http://www.total.com/MEDIAS/MEDIAS_INFOS/2151/FR/girassol-VA.pdf.

Wembridge, M., and S. Pfeifer. 2011. "BP Wins $250M for Macondo Fund from Equipment Maker." *The Financial Times*, 17 December.

Wright, R. 2010. "Oil Tankers Find Silver Lining in Spill." *The Financial Times*, 15 June.

4 Case Studies: IOC & OSC Relationship in Selected Sectors

IOCs are system integrators that sit at the centre of a complex industry structure. The relationship between the IOCs and service contractors is multilayered and complex and varies by sector. IOCs orchestrate various alliances and contractual relationships involving oil services companies, their suppliers and other joint venture operating companies with the aim of producing oil whilst reducing overall costs and ensuring access to required inputs and crucial technology (McKinsey 1997). The examples explored below aim to demonstrate the relationship between IOCs and service contractors in three different service sectors, namely offshore drilling services, exploration services and well services.[1]

Each sector illustrates company boundaries and liabilities from different perspectives. Offshore drilling services is a capital-intensive sector that is outsourced by nearly all IOCs. Well services is also outsourced to a very large extent. This sector has given rise to the largest oil services companies such as Schlumberger and Halliburton. In contrast, exploration services is a sensitive sector where IOCs are determined to keep their essential expertise of reservoir knowledge and hence outsource only certain parts. Within each case study, a general presentation of the sector and the main actors will be followed by analysis of the relationship between the IOC and the contractors.

4.1 OFFSHORE DRILLING SERVICES

'Drilling is thrilling.'

Operators need to drill exploratory wells in order to gather reserve information or development wells in order to prepare the discovery for production. The operator of the field (here, the IOC) conducts seismic surveys and studies with geologists and geophysicists to determine the potential for oil reserves. If the geological analysis offers clues indicating high probability of reserves, the operator will hire a drilling contractor to drill exploratory ('wildcat') wells. This is because drilling a well is the best way to gain a full understanding of subsurface geology and determine the potential of an oil formation. Exploratory drilling consists of digging into the earth's crust to allow geologists to study the composition of the underground rock layers in detail. Once the well is drilled, well logging[2] yields data on rock types that are present and the fluid that the rocks contain (Natural Gas 2011). The information interpreted from the logs is used to decide whether a well should be completed to produce oil or filled with cement and abandoned.

Following exploration wells, appraisal and development wells are drilled in proven areas for the production of oil. Drilling an exploration well is an expensive and time-consuming undertaking. In addition, most exploration wells turn out to be dry holes. Therefore, they are only drilled in areas where seismic data has indicated a high probability of petroleum formations. The decision to drill development wells is taken following the assessment of the reservoir with exploratory wells (Natural Gas 2011).

Non-oil-industry observers often take the ability to drill a hole down a thousand metres with a drilling rig[3] for granted. Advances in equipment and techniques, such as rotary drilling, have improved the success rates in reaching targets. Most industry observers recognize that a land drilling rig is a relatively simple and commoditized machine. However drilling in the sea is still very complicated, mainly due to the lack of stability (for floaters), the corrosive environment and difficult support logistics. Offshore fields require complex structures to drill a well to explore as well as to produce oil. The shape of these complex structures (offshore rigs and platforms) is determined largely by the depth of water and can be in the form either of jack-ups or floaters[4] (drillships and semi-submersibles) (Hermann et al. 2010). In terms of numbers, jack-up rigs drill most offshore wells, semi-submersibles come second and drillships come in third (Diamond Offshore 2012). Different types of oil rigs involve drillships, semisubmersibles, jack-ups, submersibles and land rigs (Schlumberger 2012c). Drilling units are designed and constructed by building yards and owned by specialized companies (contract drillers).[5]

Contract drillers are companies that help their customers find and develop oil reserves by providing onshore and offshore contract drilling services. They drill and complete[6] wells at the direction of their customers, which can be IOCs, independents or national oil companies.

Drilling contractors own and/or operate a fleet of onshore and offshore drilling units which IOCs contract to drill the wells. There are a small number of cases where drilling companies operate drilling in customer-owned deepwater rigs, but in most cases contractors are the owners of the rigs. The rigs can be hired on a short- or long-term basis. The contract duration usually varies from two weeks to a couple of years depending on the complexity of the project. The typical working duration for the more than 500 offshore working drilling rigs in the world is less than a year. During the duration of drilling, contractors work on a daily fee basis. Despite using similar drilling technologies, daily drilling rates are higher for offshore fields than onshore fields due to the additional complexities of the offshore environment.

Table 4.1 Example of a Drilling Fleet and Daily Rates by a Contractor

Rig Type / Name	Generation / Type	Built	Water Depth (1,000 feet)	Drilling Depth (1,000 feet)	Location	Client	Start	Expire	Dayrate (1,000 USD)	Previous Dayrate (1,000 USD)
West Alpha	4th-HE	1986	2	23	Norway	ExxonMobil	Jan-14	Jul-16	506	479
West Capella	Ultra Deepwater	2008	10	35	Nigeria	ExxonMobil	Apr-14	Apr-17	627.5	562
West Sirus	6th-BE	2008	10	35	USA	BP	Jul-14	Jul-19	535	490
West Phoenix	6th-HE	2008	10	30	UK	Total	Jan-12	Sep-15	458	544
West Eminence	6th-HE	2009	10	30	Brazil	Petrobas	Jul-09	Jul-15	609	–

Source: Compiled by the author based on data available from Sea Drill (2015).

Contractors regularly announce day rates for new rig contracts. Hence there is a regular stream of announcements of new contracts with new rates. These rates are tracked by industry observers and vary according to the rig type and the region of utilization, as well as increasing and decreasing in line with supply and demand in the rig market. As of December 2012, the daily rates range from between $253,000 to $467,000 for floaters and between $40,000 and $154,000 for jack-ups compared to an average of $20,000 for land rigs (Rigzone 2012d). The high oil prices in evidence before the 2008 crisis increased the demand for all rigs and drove prices up to record levels. At their peak in 2008, daily rates reached $700,000, largely owing to offshore drilling rig utilization rates reaching 100% around the globe (Malek et al. 2009). Rental rates for the industry's ultra-deepwater rigs, the world's most complex and expensive drilling vessels, would climb 28% to a record $714,000 a day by the third quarter of 2012, up from about $560,000 in March of the same year, according to estimates by Ole Slorer, an analyst at Morgan Stanley who dubs the move a 'super spike' (Wethe 2012).

Rig utilization rates give a clear picture of the supply–demand balance on the rig market, with rig rates going up when utilization rates are high. As of May 2015, the global fleet utilization rate was at 86.6%, with 673 rigs contracted out of a global rig supply of 860 offshore rigs (IHS 2015). Demand for rigs moves in line with new exploration or development projects by oil companies. Supply, on the other hand, changes with retirement and/or construction of new rigs. Rigs are constructed in building yards belonging to companies such as Hyundai, Samsung, Daewoo and Coco Nantong, while equipment is provided by companies like National Oilwell Varco, Cameron International and Aker Solutions. The figure below shows the link between

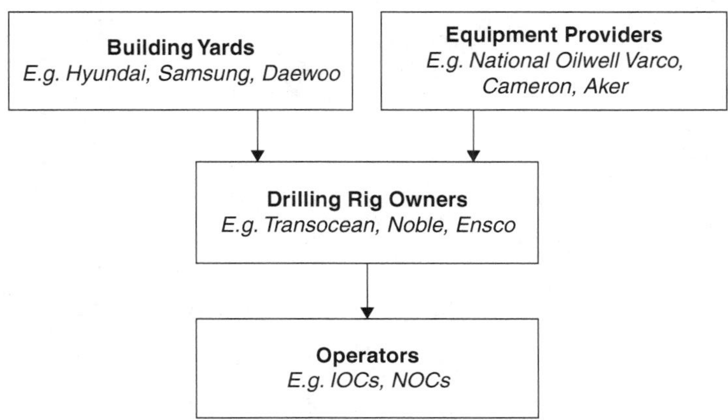

Figure 4.1 Link between Building Yards, Equipment Providers, Drillers and Operators

Source: Compiled by the author.

Table 4.2 Example of Floater Orders as of 2011

Order Date	Rig Name	Contractor	Rig Type	Rig Water Depth (1,000 feet)	Construction Status	Delivery Date	Rig Design	Build Yard	Build Cost (million USD)	First Operator
11-Nov-2010	Queiroz Delba Drsh Tbn1	Queiroz Galvao / Delba	Drillship	10	Under Construction	01-Jul-2012	Samsung 10000	Samsung Heavy Industries		Petrobas
11-Nov-2010	West Auriga	Seadrill	Drillship	12	On Order	01-Mar-2013	Samsung 10000	Samsung Heavy Industries	595	
19-Jan 2011	Noble Drsh Tbn5	Noble	Drillship	12	On Order	01-Jul-2013	GustoMSC P10000	Hyundai Heavy Industries	605	Shell
20-Jan-2011	Aker Drsh Tbn2	Aker Drilling	Drillship	12	Planned	01-Dec-2013		Daewoo	600	
02-Feb-2011	Ocean BlackHornet	Diamond Offshore	Drillship	12	On Order	01-Dec-2013	GustoMSC P10000	Hyundai Heavy Industries	590	

Source: A section of the table 'Current Cycle Floater Orders' appearing in West et al. (2011).

companies connected to drilling, whereas the following table provides an example of new orders.

Rig contractors aim to increase their fleet via new construction or acquisitions of other rig companies.

Market Structure and Main Actors

In the rig market, drilling companies may be ranked by different criteria such as market capitalization or profitability. Within the context of this book, a ranking based on the number of rigs seems appropriate because the contractor's fleet size and the capabilities of its rigs impact the drilling company's negotiation power with the IOCs.

The number of rigs in each contractor's fleet changes every year with the addition of newly constructed rigs and the retirement of old rigs. The number of mobile offshore drilling units (MODUs) in the fleet, the number of rigs working and rigs under construction all vary on an annual basis. As of 2012, the biggest offshore drilling contractor is Transocean, followed by Noble and Ensco.

The traditionally fragmented industry has been consolidated in recent years, and the ranking[7] of companies on the list has changed considerably.

Table 4.3 The Biggest Contractors by Rig Fleet as of 2012

Contractor	Rig Fleet	Rigs Contracted	% Utilization	Headquarters
Transocean	138 rigs	99 rigs	71.70%	US
Noble Drilling	82 rigs	59 rigs	72.00%	US
ENSCO	77 rigs	64 rigs	83.10%	UK
Nabors Offshore*	71 rigs	21 rigs	29.60%	US
Hercules Offshore	64 rigs	25 rigs	39.10%	US
Seadrill Ltd	57 rigs	38 rigs	66.70%	Norway
KCA Deutag*	45 rigs	41 rigs	91.10%	UK
Diamond Offshore	44 rigs	31 rigs	70.50%	US
China Oilfield Services Ltd.	34 rigs	26 rigs	76.50%	China
Maersk Drilling	34 rigs	24 rigs	70.60%	Denmark
Rowan	34 rigs	25 rigs	73.50%	US
Seawell	30 rigs	30 rigs	100.00%	Norway
Sete Brasil S.A.	29 rigs	0 rigs	0.00%	Brazil
PDVSA (NOC)	26 rigs	8 rigs	30.80%	Venezuela

Source: Compiled by the author based on data available from Rigzone (2012e).

*Note: *Because the fleets owned by Nabors Offshore and KCA DEUTAG mainly consist of platform rigs (a stationary offshore oil and/or gas production facility), these companies are not included in the list of top drilling contractors by* Offshore *magazine in Table 4.5.*

Historically, the offshore drilling sector was fragmented with a wide range of companies building and operating floaters. In the early 1970s, new floaters such as semi-submersibles and drill ships with new depth capacities (3,000 feet) were developed. By the late '70s, the drilling sector had become highly disjointed with a large number of companies operating these floaters. Oil services companies invested heavily in equipment and rigs following a drilling boom in the early 1980s when oil prices reached a peak of $35 per barrel. When the price of oil fell to below $10 per barrel by 1986, oil companies cancelled their drilling programmes or negotiated lower rates, causing several drilling companies to go bankrupt or be acquired by stronger rivals. During the decade-long lean period, the number of offshore drilling rigs in operation declined swiftly, from more than 1,000 in the early 1980s to around 500 a decade later. The industry remained fragmented. By 1995, the top three companies served just 27% of the market. Four hundred jack-up rigs owned by eighty separate companies created a supply–demand imbalance in favour of oil companies over drilling contractors. Companies were not able to ensure long-term success by expanding their business through the purchase of new rigs. They concluded that growth had to come by acquiring existing rigs through consolidation with other contractors and hence by gaining leverage with oil companies (Funding Universe 2012).

1n 1999, Transocean started to emerge as the leading player with a significant market share when it bought Schlumberger's spin-off Sedco Forex. Following the merger, Schlumberger chairman and then CEO Euan Baird stated that 'the drilling industry has, as a whole, been characterized by volatile earnings and weak balance sheets. Today, as the industry goes to deeper water and harsher conditions, with even higher investments in equipment and technology needed to meet customer requirements that now include global sourcing of services, the structure of the industry has to change'. He called the merger 'the first decisive move to create industry leadership' (OGJ 1999). At the time of the merger and on a combined basis of floater fleet capacity, Transocean Sedco Forex held 24 to 25% of the market while its nearest competitor held only 10 to 15% (Offshore 1999).

Transocean continued to consolidate its position, merging with R&B Falcon in 2000 and with its next largest competitor, GlobalSantaFe, in 2007. The combined company, under the name of Transocean, continues to be the market leader with a fleet of nearly 150 rigs. Although the position of Transocean and Noble at the top of the rankings has not changed, several acquisitions in the past five years, especially during 2010 and 2011, have changed the rankings within the list of the top ten drilling contractors. Four out of the top six contractors have increased their rig fleet through acquisitions.

The table below comparing the top ten contractors in 2007 to 2012 demonstrates the consolidation that has taken place among drilling contractors. Companies such as Seadrill, Hercules Offshore, Noble and Ensco have increased their fleet and elevated their market share through acquisitions.

Table 4.4 Recent Offshore Rig Company Consolidation

Buyer	Seller	Date	Value Implied by Acquisition	Value per Rig	Jack-ups / Floaters
Rowan Companies	Skeie Drilling	Q3 2010	$1.2 billion	$410 million	Jack-ups
Seadrill	Scorpion Offshore	Q3 2010	$1.35 billion	$193 million	Jack-ups
Noble	Frontier Drilling	Q3 2010	$2,16 billion	$309 million	Floaters
Hercules Offshore	Seahawk Drilling	Q1 2011	$106 million	$5 million	Jack-ups
Ensco	Pride International	Q1 2011	$8.6 billion	$430 million	Both

Source: Information based on the table 'Recent Offshore Rig Consolidation Activity' in West et al. (2011).

Table 4.5 Rig Fleet of Top Drilling Contractors from 2007 to 2012

2007		2008		2009		2010		2011		2012	
Transocean	81	Transocean	142	Transocean	140	Transocean	141	Transocean	139	Transocean	142
Global SantaFe	60	Noble	61	Noble	61	Noble	62	Noble	69	Noble	79
Noble	60	Ensco	49	Ensco	51	Ensco	50	Seadrill	56	Ensco	76
Ensco	48	Pride	47	Pride	46	Diamond O.	46	Ensco	48	Seadrill	59
Pride	48	Diamond O.	45	Diamond O.	45	COSL	33	Diamond O.	47	Hercules O.	51
Diamond O.	47	SeaDrill	37	SeaDrill	43	Hercules O.	33	Hercules O.	34	Diamond O.	50
SeaDrill	34	Hercules O.	36	Hercules O.	38	Maersk	29	COSL	32	Rowan	34
Maersk	30	Maersk	32	Maersk	30	Rowan	28	Rowan	31	COSL	33
PDVSA	28	Rowan	30	Rowan	29	PDVSA	25	PDVSA	29	PDVSA	29
Todco	27	PDVSA	27	PDVSA	27	Pride	25	Pride	26	Maersk	28
TOTAL	463	TOTAL	506	TOTAL	510	TOTAL	472	TOTAL	511	TOTAL	581

Source: Compiled by the author based on data published each year by Offshore (Offshore 2007, 2008, 2009, 2010, 2011a, 2012).

The principal advantage of acquiring another rig company is the increase in the size and range of the rig contractor's fleet. A wide range of rigs contracted at different time maturities helps rig owners increase flexibility in their portfolio and lessens their exposure to changes in day rates. In addition, acquisitions have been justified to ensure a global presence in order to decrease the costs of rig mobilization and build up a diversified customer base. This is illustrated in the CEO's statement during the Transocean–GlobalSantaFe merger in 2007. GlobalSantaFe President and CEO Jon A. Marshall stated:

> This is an exciting opportunity for our shareholders, our customers and our people. The $15 billion cash payment allows us to achieve a more appropriate capital structure and deliver immediate value to our combined shareholders. The combined company will have a broader customer base, particularly with the increasingly important national oil companies, greater exposure to the growing deepwater business and increased low-risk organic growth prospects from the combined deepwater new build programme. The enhanced operational capability of a more geographically diverse rig fleet will produce significant benefits for our customers and provide substantial growth opportunities for our people. This is an ideal fit for the stakeholders in both companies.
>
> (Transocean 2007)

Smaller companies which do not have a large amount of geographic dispersion may be excluded from bidding on certain oil projects because moving a rig from a distant location could result in costly mobilization (Offshore 1999).

Following the BP accident, additional costs have also been cited among the reasons for consolidation. The costs of additional safety and environmental regulations, alongside the quality improvements required in the post-Macondo world, make critical mass even more important and favour large companies. The CEO of Pride International, Louis Raspino, said that 'in general, in a post-Macondo world, increasing critical mass is more important than in a pre-Macondo world. Larger companies are considered to have greater ability to shoulder the costs of additional regulation and quality improvements required around the world, as well as an edge in retaining the best people' (Clanton 2010).

Finally, consolidation is also a result of the sophistication required to go to deeper and more technically challenging areas. If oil was still to be found in areas that require only simple technologies such as the pump jacks used on onshore fields (also called 'nodding donkeys'), there would continue to be lots of small companies and no concentration would have been necessary.

Further rig company consolidation is expected by market observers because limited near-term shipyard availability and rising construction costs will make it difficult for traditional drillers to increase their fleet through the construction of new rigs. According to Barclays, ageing offshore equipment that is near retirement age, the need for higher quality rigs and increased focus on safety in the post-Macondo era have been driving the new rig-construction cycle. For those who have not participated in the new rig-construction cycle, the only other growth option is via the acquisition of another drilling contractor, an option which may also prove more economical than the purchase of new rigs (West et al. 2011).

Current top offshore drilling contractors vary in their capabilities and geographical presence; most of them are active in all shallow, deep and ultra-deepwaters although a few do not have the rigs and capability for ultra-deepwaters. The majority are headquartered in the US with offices and drilling activities located throughout the world. Several interviewees confirmed that attaining market leadership in the sector requires varying capabilities, a diverse fleet with various technical specifications and a worldwide presence. As the situation demonstrated after the drilling moratorium that followed the BP accident, companies which focus their presence in one region (such as the Gulf of Mexico) are prone to facing difficulties in the case of a problem in that region. For example, Seahawk Drilling suffered financially as a result of the oil-spill-related slowdown in Gulf of Mexico drilling activities and filed for bankruptcy. Consequently, Hercules Offshore acquired its rigs in February 2011 in order to build up a larger, more diverse fleet, broader customer relationships and greater operational flexibility (Reuters 2011). Table 4.6 on page 111 demonstrates the different capabilities of drilling contractors and their country of incorporation.

The top ten firms globally are estimated to account for 70% of the total rig fleet, with Transocean possessing 17% of the global market share, while the top three, namely Transocean, Noble and Ensco, control a total of 35%. The market concentration also varies according to rig types. The market share is more concentrated in ultra-deepwater drilling and more dispersed in the jack-up rig market. Transocean owns nearly half of the fifty or so deepwater platforms in the world (Krauss 2010). The deepwater segment provides the highest profit margin and is highly consolidated because deeper areas require advanced technologies that can only be provided by a few companies.

Transocean: Transocean is the world's number one offshore drilling company. It is also the owner and operator of the $365 million Deepwater Horizon rig at the centre of US Gulf Oil Spill disaster (Wethe 2012). Today's Transocean is composed of several drilling companies that were merged.[8] Transocean ASA was created in the mid-1970s by a Norwegian whaling company entering into the semi-submersible business and later consolidating with other companies. Transocean possessed large North Sea operations and made the new company, Transocean Offshore Inc., a leader in deepwater drilling. Transocean Offshore Inc. began building large drilling operations with drillships such as *Discoverer Enterprise*[9] that could drill, test and complete wells in water depths of up to 10,000 feet.[10] In 1999, it

Table 4.6 Capabilities of Drilling Contractors

Exploration & Appraisal Drilling Services		Onshore	Shallow Water	Deepwater	Ultra-Deepwater
Europe	Abbot Group	✓	✓		
	Ensco	✓	✓	✓	✓
	Fred Olsen		✓	✓	✓
	Maersk		✓	✓	
	Saipem	✓	✓	✓	
	Seadrill		✓	✓	✓
US	Diamond Offshore	✓	✓	✓	✓
	Helmerich Payne	✓		✓	✓
	Hercules Offshore		✓	✓	
	Nabors	✓	✓	✓	✓
	Noble Drilling		✓	✓	✓
	Precision Drilling	✓			
	Rowan		✓	✓	
	Transocean		✓	✓	✓
Asia	China Oilfield Services	✓	✓	✓	✓

Source: Compiled by the author based on information in Hermann et al. (2010, 78–79).

merged with Sedco Forex,[11] a Schlumberger offshore contract drilling operations spin-off, on an equal basis (a merger of equals) to create the world's largest offshore drilling company, Transocean Sedco Forex. Following the acquisition of R&B Falcon[12] in 2000, Transocean Sedco Forex became the third-largest oil services company after Schlumberger and Halliburton, with a company valuation of around $14 billion. In 2003 the company simplified its name to Transocean, and in 2007, Transocean, the world's biggest drilling company, and Global SantaFe,[13] its nearest rival, agreed to a merger of equals. The combined company will have a global fleet of 146 rigs, including harsh-environment jack-ups and ultra-deepwater drillships and an estimated value of $53 billion including debt (NBC News 2007). The deal combined the world's two largest drillers in a market where demand for rigs has never been higher[14] (NY Times 2007). The new company is triple the size of its largest peer, Noble Corp., in sales with an order backlog of $33 billion (Strahan 2007). Transocean has recently continued to increase its fleet by buying other companies. In 2011, it acquired the Norwegian ultra-deepwater drilling company Aker Drilling and upgraded its fleet by four rigs in the tightening ultra-deepwater market segment. The purchase valued the four ultra-deepwater rigs at $797 million per rig, compared to replacement estimates of $807 million according to Terra Market's analyst (Wethe and Stigset 2011).

Transocean has taken the lead in the consolidation of offshore drilling contractors. In 2010 a number of contractors merged, but no one has come close to rivalling Transocean in size, especially in the deepwater and harsh environment offshore drilling markets where the company leads the field with its twenty-seven ultra-deepwater and sixteen deepwater floaters. Although Transocean's fleet includes swamp barges, shallow water jack-ups and mid-water floaters, the company is especially active in the deepwater and harsh environment drilling segment, offering semi-submersible rigs as well as massive drillships that have drilled to record depths in the range of 10,000 feet. Transocean holds technical leadership in the sector. Its technical achievements have progressed over time from building the first offshore jack-up drilling rig in 1954 to developing the first ultra-deepwater drillship with patented dual-activity drilling system and the first drillship capable of working in 10,000 feet (3,048 metres) of water. It holds approximately 80% of the world records for drilling in the deepest waters, including 10,194 feet (3,107 metres) of water offshore from India by the ultra-deepwater drillship, and the deepest well ever drilled offshore at 35,050 feet (10,683 metres) by the semi-submersible rig Deepwater Horizon while working for BP in the US Gulf of Mexico. Its mobile rigs cover all of the world's major offshore drilling markets, including the Gulf of Mexico, the North Sea, the Mediterranean Sea and the waters off eastern Canada, Brazil, West and South Africa, the Middle East, Asia and India. The company holds the highest market share in the overall drilling market and nearly 50% of the ultra-deepwater market. The average day rate of Transocean moved from $211,900 in 2007 to $283,800 in the third quarter of 2009 (Funding Universe 2012; Transocean 2011, 2012).

Noble: Noble was founded in 1921 as a one-rig operation and has since developed into the second-largest offshore drilling contractor. Noble Drilling Company was part of Noble Corporation until its spin-off in 1985. Noble Drilling's growth since the spin-off has come through the acquisitions of offshore drilling assets and non-capital intensive businesses, Noble's aim being to gain a strong position in foreign markets, to expand marine drilling operations and to move into new markets and segments of the industry. Noble acquired drilling rigs, purchased drilling companies,[15] sold certain business such as land-drilling assets[16] and formed JVs[17] throughout its history in order to achieve its current size. In its most recent deal, Noble bought Norway's Frontier drilling for $2.16 billion in 2010, adding seven vessels to its fleet. In addition to five vessels under construction, Frontier, which is domiciled in Norway and headquartered in Houston, has been building two ultra-deepwater drillships designed to operate in Arctic waters under a partnership with Shell (Daily 2010). Today, Noble provides offshore drilling services in major basins of the world with its fleet of seventy-nine offshore drilling units (including five ultra-deepwater and six jack-up drilling rigs) (Noble 2011, 2012).

Ensco: Ensco is a relatively new company. It was incorporated as Blocker Energy Corporation in the US in 1975. It started as Energy Services

Company (Ensco) in 1987 and continued growing through the acquisition of Penrod Drilling (1993) and Dual Drilling (1996). In its first ten years, from 1987 to 1996, Ensco purchased and refurbished equipment and branched out from its contract drilling business into various associated businesses, including a tool and supply company, engineering services and a marine transportation business. During the next decade, in the years between 1997 and 2008, Ensco changed its strategy and decided to focus solely on offshore drilling with a premium fleet. It divested itself of marine vessels, platform rigs and the majority of barge rigs, aiming to grow its jack-up fleet and enter the ultra-deepwater market with the delivery of its first semi-submersible in 2000. In 2002, it acquired Chiles Offshore, which owned four high-specification jack-ups. In 2003, it exited the marine transportation business completely by divesting its fleet of support vessels. Ensco has continued to follow the same strategy, focusing on ultra-deepwater and premium jack-up drilling markets, to the present date. It moved its headquarters to the UK in 2009 and made a large acquisition of Pride International in May 2011 for $7.3 bn to create the second-largest offshore driller in the world. The combined company will have seventy-four rigs including twenty-one ultra-deepwater and deepwater rigs, forming the second-largest and youngest fleet able to drill in water depths of 4,500m or greater. The acquisition provided Ensco with access to the fast-growing deepwater markets of Brazil and West Africa and diversified its asset base from consisting mainly of jack-up rigs to drillships and semi-submersible rigs. According to the press release following the merger, the estimated enterprise value of the combined company is $16 billion (as of February 4, 2011), whereas the total estimated revenue backlog for the combined company is approximately $10 billion (Ensco 2011b). Today Ensco provides drilling management for three customer-owned deepwater rigs and has its own fleet of seventy-six rigs consisting of dynamically positioned drill ships and semi-submersibles, moored semi-submersibles and jack-ups. Ensco owns and operates one of the world's newest[18] ultra-deepwater fleets and largest fleets of active premium jack-ups. It has presence in most strategic offshore basins across six continents (Ensco 2011a, 2012).

Relationship with International Oil Companies

Offshore drilling activity is outsourced to drilling contractors by nearly all IOCs. This is mainly due to the combination of the high cost of drilling rig ownership and the intermittent nature of drilling activities. Regarding cost of construction, rigs can vary widely in price. It has been reported that the Deepwater Horizon rig at the centre of the BP accident cost $350 million to build in 2001 (National Commission 2011). However, this reported cost seems lower than is the case for most rig construction, which tends to vary between $175 and $225 million for jack-ups and $500 and $700 million for offshore floating rigs (Kaiser and Snyder 2012). Rig contractors order new rigs either on a speculative basis foreseeing a surge in upcoming activities or in advance of work arranged for their customers (Drilling Contractor 2002). Because the drilling schedule of IOCs varies greatly from year to year and is dependent on several factors, such as oil prices or the results of seismic studies, it does not justify the ownership of a drilling rig on a continuous basis. The capital-intensive nature of building up a fleet with rigs adaptable to the different sorts of well that an oil company might be operating is not a good fit with the intermittent nature of the drilling activities undertaken by IOCs. Interviewees mentioned that while in the past some of the IOCs used to own rigs, they have now sold them to service companies.

The relationship between drilling contractors and IOCs is set up on a hiring contract basis. IOCs hire offshore rigs with a full crew and the equipment and supplies needed to drill the assigned well. Offshore rigs are self-sufficient units with crew living on board and typically working on a rotating basis (several weeks on, several weeks off). These rigs are designed for efficiency in living and working, with emphasis on keeping the rig steady in gulf or oceanic waters.

According to interviewees, the sharing of tasks on the rig is clearly defined. All decisions related to the well are taken by the IOCs, whereas decisions related to the rig (such as working conditions, organization of support vessels, etc.) are taken by the rig contractor. IOCs have their own staff on the rig supervising the drilling work. Decisions made on 'the well' belong to the supervisor. Day-to-day management of the drilling remains the remit of the IOC, with daily instructions passed to the drilling contractor by the IOC.

The sharing of the tasks is reflected in the allocation of liabilities. The clear allocation of liabilities on an offshore rig is necessary for two main reasons. First, the operations of drilling rigs are potentially very hazardous because rigs hold large quantities of gas and liquid hydrocarbon material, which is explosive, flammable and potentially highly dangerous. The recent BP accident during exploratory drilling is a prime example of how drilling a well can go wrong. Second, the work entailed in the well involves many contractors and subcontractors. Without clearly defined roles, functions and areas of responsibility between all involved parties, companies cannot invest large capital sums in technically risky projects in difficult environments. Therefore most drilling rigs and well services contracts have special clauses allocating liability owing to the highly risky nature of operations. Most drilling contracts share liabilities under a 'mutual hold harmless' or 'knock-for-knock' indemnities regime. Under such a regime, 'each party to the contract agrees to take responsibility for, and to indemnify the other against, injury and loss to its own personnel and property and its own "consequential losses" (by this, the parties generally mean loss of profit and other economic losses). These cross-indemnities are usually intended to be effective even if the losses arose due to negligence, breach of statutory duty or breach of contract' (Hewitt 2008, 6).

To clarify, in the example of a drilling contract that contains a true knock-for-knock indemnity clause, both the IOC and the drilling contractor agree to indemnify the other against claims for injuries or damages that they or their employees sustain regardless of who was at fault or who was negligent. Therefore, if there is any problem with the rig itself, the drilling contractor to whom it belongs is responsible. The contractor would protect the IOC and its employees against claims and would be responsible to defend and pay any damages awarded regarding the rig. It follows that the contractor would be responsible for the removal of the wreck and the construction cost of a new rig in case of a blowout accident.

On the other hand, the well or the reserve is the responsibility of the operator. Even in the case of gross negligence on the part of the drilling contractor, the IOC is responsible to indemnify for any damages to the IOC's property or employees and has to indemnify the drilling company against any claims and pay any damages awarded. Therefore for a typical drilling contract, subject to the drafting of the contract in question and to the respective commercial bargaining positions of the parties, pollution emanating from the reservoir, loss of or damage to property or equipment while 'downhole', loss of or damage to the well, blowout, fire, explosion, or any uncontrolled well condition, damage to the reservoir or geological formation, etc., are borne by the IOC. As such, each of the parties insures its own goods, holding other party harmless for any damage regardless of who was at fault or who was negligent. There are contracts where parties have excluded gross negligence or wilful misconduct from these indemnities, but in general wilful misconduct remains the only exception and the mutual harmless principle rules even under gross negligence. As a summary, the offshore oil industry has developed liability allocation models to deal with the huge risks involved as well as capital sums invested and lives at stake (Hewitt 2008).

A clear allocation of liability is not the only component of a relationship between the IOC and the drilling contractor, since the risks in offshore drilling are not limited to the occurrence of an accident. Among all the undertakings in the world, the industry experts hold oil drilling to be one of the riskiest and most expensive. Although advanced seismic studies can reduce risk, the possibility of finding a dry hole and losing huge sums of money cannot be excluded. The cost of drilling can be significant to IOCs. The daily rate for an ultra-deepwater rig is subject to constant fluctuation but can reach $500,000/day to contract, and $1 million per day to operate including helicopters, support vessels and other services (Rigzone 2012d).

Mukluk Island illustrates industry risks and potential losses. It has been reported that in 1982–83, twelve companies spent nearly $2 billion on drilling in the Beaufort Sea, north of Alaska. The well turned out to be a dry hole and has since gained notoriety as the most expensive dry hole in oil industry history (Udall and Andrew 2008). In addition to risk of not finding oil, drilling operations may face delays due to problems with equipment or weather conditions. For example, Shell's drilling operations in the Chukchi Sea (Arctic) have faced a series of setbacks and delays due to ice floes crushing equipment and problems caused by extreme weather conditions (Reuters 2012). In the case of such delays or a moratorium due to an accident, parties will follow the existing provisions in the contract if the case has already been envisaged, or renegotiate the terms. For example, following the BP accident, Noble said it had resolved the status of its units contracted to Shell in the Gulf of Mexico and would allow the oil giant to suspend contracts at a reduced rate. Noble has stated that the contracts will resume at their full terms and normal rates once drilling resumes. According to Argus Research analyst Phil Weiss, Noble also negotiated a three-year extension on its Noble Jim Thompson rig currently operating in the Gulf at a reduced rate of $336,200 per day, a sum significantly lower than the $400,000 to $500,000 daily rate that rigs were bringing in before the spill (Daily 2010).

Regarding negotiating power, some interviewees argued that consolidation in the drilling market has reduced the leverage of the IOCs. In the wake of Transocean Sedco Forex's bid to buy R&B Falcon, Bob Rose, the CEO of Global Marine, argued that due to their tremendous size, international oil companies always have advantage when it comes to contract negotiations. He said:

> If a drilling company tries to negotiate an operator into a corner, that operator can easily back a competing driller, offering contracts that would allow that driller to build new rigs to service the demands of that operator's drilling programme. If you are a $200 billion E&P company, a $15 billion drilling contractor really is no threat at all. For example, at one time Diamond, Global, and Transocean were all trying to hold new build deepwater rates at a high level. Instead, Exxon turned to Marine Drilling and committed to contracts that allowed the company to build its first deepwater rig. All of these oil companies are capable of being kingmakers, if they choose to do so.
>
> (Bob Rose quoted in Offshore 2000)

On the other hand, other industry experts affirm that consolidation in the drilling market has reduced the negotiating power of IOCs, as is evidenced by the huge rise in daily drilling rates in the ultra-deepwater market. In order to circumvent the high daily drilling rates and the scarcity of rigs in the ultra-deepwater market, IOCs are attempting to bind OSCs into longer term contracts with clauses allowing them to extend for a further period; Statoil, in particular, has signed multiple large rig contracts with primary terms of four to five years with up to five additional years as an option (Offshore 2007).

Another indicator of deep relationship is the consideration of IOCs' requirements in long-term planning by drilling contractors. Oil services companies order the construction of new rigs in anticipation of employing these rigs specifically for one company, or generally on the market. For example, on its webpage Transocean states that as part of the hedging tactics practised by IOCs, both Exxon USA and Chevron Oil had each committed to five-year contracts for a new generation of drillship. As a result, incorporating the most up-to-date features of naval architecture, marine engineering and floating drilling technology, the Glomar Pacific was delivered in mid-1977 and the Glomar Atlantic in 1978 to these companies. In a more recent example, in 2006 Noble announced the receipt of a letter of award from Venture Production for a new F&G JU-2000E enhanced premium jack-up drilling rig to be constructed in China. Venture Production informed the company that it anticipates employing the rig in the North Sea. Noble also received a commitment from Shell Exploration and Production for the upgrade of the Noble Clyde Boudreaux.

In most cases, the specifications of new rigs are given in line with IOCs' requirements. Ensco states that the ENSCO 8500 Series ultra-deepwater semi-submersible rig, capable of drilling in up to 8,500 feet of water, uses a proprietary design developed with extensive input from customers to address the drilling requirements of nearly every deepwater field around the world. These rigs have built a history of outstanding performance and continue to attract repeat customers (Ensco 2011a).

This is particularly important because the design of the well, the choice of the drilling site and the choice of the type of drilling are decided by IOCs. Producing an optimal well design and drilling programme is a long and complex process, involving a wide range of people with different specializations and priorities. It needs to be managed with a similar level of attention and accuracy as is required in drilling the well itself. Poor planning can escalate cost enormously, and the root causes can be difficult to pinpoint. The oil company chooses the location and selects the rigs that are specifically suited for a particular job, because each rig and each well has its own specifications and the rig must be matched to the well (Diamond Offshore 2012).

4.2 SEISMIC EXPLORATION SERVICES

'Exploration begins with seismic.'

Oil reservoirs are located below the earth's surface and hence their presence and size below ground cannot be estimated visually. The existence of oil in formations beneath the earth's surface can only be determined by drilling. Drilling provides companies with information regarding the presence as well as the characteristics of the reservoir. The essential question to be answered before any drilling activity is where to drill. Exploration work is thus required. Because drilling is costly, oil companies need to drill at locations with a high probability of oil presence. It is therefore vital to conduct seismic exploration in order to determine the highly probable areas for oil-bearing formations. Seismic exploration gives a picture of the earth's subsurface, which assists oil companies to look beneath the surface and locate oil and gas reserves. Its main purpose is to render the most accurate possible graphic representation of specific portions of the earth's subsurface geologic structure (CGGVeritas 2012c). Seismic operations perform this function by using sound waves to create an image of subsurface rock layers.

> The principle is the same as a dolphin's sonar or a medical scan: an acoustic wave (seismic wave) is sent into the ground, where it is reflected by each interface between two geological layers and ultimately recorded at the surface by a receiver. The time it takes for the wave to travel from the surface to the geological interface and back will give an idea of the depth of the layer, provided the wave propagation speed is known. Cumulating the data gives an idea of the whole geological structure of the subsurface.
>
> (Total 2012, 58)

Sound waves, generated by a loud 'bang', propagate down through the earth and are partially reflected by each rock strata boundary back to the surface. All reflected waves are recorded by sensors (geophones or hydrophones) at the surface and the information is then digitized and stored in tapes (Hermann et al. 2010). Measuring devices measure the strength of each reflected wave as well as the time it has taken for the wave to travel through various layers of the earth's crust and back to the location of the sensors (OGP and IAGC 2011).

Seismic exploration involves acquisition, processing and interpretation of seismic data. Both on land and at sea,[19] the acquisition of seismic data involves the transmission of controlled acoustic energy (an output pulse, a 'bang') into the earth, and recording the energy that is reflected back from geologic boundaries in the subsurface. 'Reflected seismic response is a mixture of our output pulse, the effect of the Earth upon that pulse, and background noise, all convolved together' (CGGVeritas 2012c). Processing of the acquired data implicates removal of the output pulse and background noise in order to produce the best possible image of the subsurface geological structure. Processing involves pre-processing including noise suppression, velocity analysis and velocity model building, migration and ensuring the fidelity of seismic amplitude and phase. It is about making the seismic data understandable. Millions of data are passed through mathematical treatment using diverse algorithms and organized in a coherent way in order to create 2D and 3D images. Measurements in time (seismic) are converted to seismic images through the data processing, which requires extensive computer power: 'For example, the amount of seismic data recorded by CGGVeritas during just one medium-sized marine 3D survey would fill more than 20,000 compact disks, forming a stack over 650 feet high' (CGGVeritas 2012c).

Following the processing, the interpretation of the processed data and integration of other geo-scientific information are carried out to assess the likelihood of oil formation and to decide the drilling location. The resulting processed data is interpreted by geophysicists and geologists. The interpretation of geological data by the experts can vary drastically; the correct interpretation is a critical step in avoiding dry holes.

In summary, seismic data is collected in the field, processed with highly advanced computers and interpreted by geophysicists. The end product of seismic work and technology is a graphic 3D[20] representation of the earth's subsurface geologic structure (Hermann et al. 2010; Hill 2011). Based on this information, exploration companies will decide whether and where to drill for oil and gas. Because accurate seismic images can mean the difference between success and an expensive dry hole, the seismic sector is critical to drilling success (CGGVeritas 2012c).

Seismic exploration is provided by geophysical seismic companies. The industry gathers, interprets and maps geophysical data. Services and equipment offered by geophysical companies include seismic data acquisition, data processing and interpretation, data management, software solutions, geological and geophysical services and seismic equipment, such as geophones or hydrophones. Companies might be focused exclusively on one of these services or offer a wide range of geophysical services.

Large seismic companies offer services in four main areas:

- Land-based and marine-based data acquisition for specific clients;
- Multi-client data acquisition that the seismic company keeps in a database and licences to a number of clients on a nonexclusive basis;
- Processing, imaging and interpretation of geophysical data; and
- Data management and reservoir studies.

A small number of companies, such as CGGVeritas and ION Geophysical, also offer seismic equipment and systems.

The majority of geophysical companies undertake both single-client and multi-client seismic surveys. Single-client work, called 'contract survey' or 'proprietary survey', can be agreed as either turnkey or term rate. Turnkey contracts are agreed either for a lump sum amount or for a fixed fee per square kilometre of data acquired. Term rate contracts provide payments based on agreed rates per units of time, typically expressed in number of days. In contract surveys, the client determines the scope and the extent of the survey, receives all the tapes with the acquired data and retains the data ownership. While small-size surveys can have a duration of one or two months and large ones two to three years (exceptionally five years), typically surveys take six months on average. For large areas, data is acquired over one year and processed during the next. On multi-client surveys, the seismic company invests in the survey, retains ownership of data and sells it to several clients on a nonexclusive basis. While multi-client surveys are particularly useful in utilizing spare capacity, contract surveys generate higher revenue for seismic companies. For CGGVeritas, marine seismic acquisition contracts accounted for 31% and multi-client marine accounted for 11% of consolidated operating revenue in 2011 (CGGVeritas 2011). For PGS, the leading marine geophysical company, single-client contracts generated 50% of its revenue, whereas multi-client library and data processing generated 40% and 10% respectively in 2011 (Reinhardsen 2012). While most companies conduct both single and multi-client surveys, there are also companies such as TGS[21] that focus exclusively on multi-client projects worldwide and only provide multi-client geosciences data (TGS 2011). UK–Norwegian company Spectrum decided to base its business on the same model of solely targeting multi-client projects by short-term chartering, and bought the 2D data held by CGGVeritas to build up its library (McBarnet 2012).

The cost of contract surveys depends on several factors including but not limited to the size and characteristics of the area, the technology required, the duration and the cost of the crew. Marine seismic surveys can cost upwards of $200,000 per day (API 2011). A 3D seismic survey may cover many square miles of land and can cost $40,000 to $100,000 per square mile or more (McFarland 2009). Whilst most contract agreements are kept confidential, an interviewee posited a cost of $3–7 million for small seismic surveys and $10–20 million for large marine ones. A handful of public announcements confirm similar figures. ION Geophysical stated a range of between €5 and €50 million in its 2010 presentation. Hyperdynamics Corporation awarded a 3D seismic contract to PGS to process the data resulting from a 3D seismic survey to be conducted over its offshore block (3,675 km^2) in the Republic of Guinea. The seismic data processing is expected to cost about $2.5 million, bringing the estimated total cost of the acquisition and processing work to $25 million (Hyperdynamics 2010).

The cost also depends on the overall supply–demand balance on the seismic services market. Demand has mirrored the price of oil. According to CGGVeritas's (2011) annual report, the offshore market experienced an excess supply between 1999 and 2004 and faced downward pressure on prices. After the downturn in 1999, major contractors, most notably WesternGeco, reduced their fleet size (McBarnet 2012). However, because of the high fixed costs in this sector, excess supply was not reduced entirely by operators but rather channelled into multi-client studies (CGGVeritas 2011). Rapid and significant price recovery occurred in 2004 and continued until the 2008 economic crisis. In 2006, with the rocketing of oil prices, various speculative marine geophysical ventures, notably Multiwave Geophysical, Wavefield Inseis, Arrow Seismic and Eastern Eco, were formed, especially in Norway where most of the marine seismic industry and operations are based. These newly formed companies had the advantage of booking shipbuilding capacity at the time of scarcity, but all have ceased to exist because they were bought by major seismic companies looking to extend and renew their fleet via acquisitions rather than building new vessels. CGGVeritas expanded and revitalized its fleet by acquiring Multiwave Geophysical (2006) and Wavefield Inseis (2009), whereas Arrow Seismic was acquired by PGS and Eastern Echo by Schlumberger in 2007 (McBarnet 2012). Decline in oil prices in 2008 and 2009 reduced demand for seismic

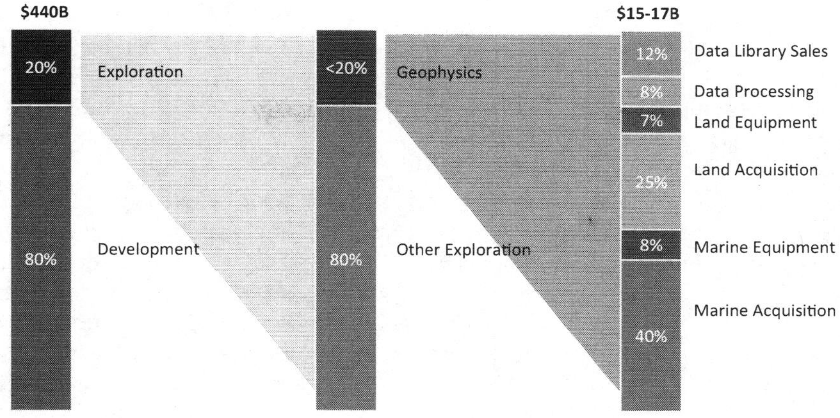

Figure 4.2 Seismic Spending versus Overall E&P Expenditures

Source: Figure based on the information in Friedemann (2010, 6).

services, and together with the increase in global offshore fleets, it produced the same downward pressure on prices. This has caused casualties, such as when Scan Geophysical & Bergen Oilfield Services (BOS) filed for bankruptcy in 2009. The BOS experience would appear to demonstrate the difficulties inherent in competing over the long term in the worldwide market for marine seismic with a small fleet: mobilization costs alone make the seismic vessels uncompetitive if any great distance to survey locations has to be travelled, and there are none of the economies of scale in crew and equipment that benefit the larger operators. These bankruptcies did not deter new companies from launching; Polarcus and Dolphin Geophysical, both Norwegian, entered the market in 2008 and 2010 respectively (McBarnet 2011). Despite growth, the marine market remained oversupplied in 2010 (CGGVeritas 2011).

Market Structure and Main Actors

According to diverse estimates, the global seismic exploration market was worth $9 to $17 billion in 2010 and around $10 billion in 2011 (Deloitte 2010; ION 2010; Visiongain 2011). According to Deloitte, the $9 billion global seismic exploration market has been split between seismic fieldwork and data processing (83%) and equipment manufacturers (17%) (Deloitte 2010). Ion Geophysical, on the other hand, estimates a $15 to $17 billion value market split between acquisition services (70%), seismic equipment (10–15%) and data processing and library (20%) (Friedemann 2010).

Seismic fieldwork and data processing can be done on land or offshore. Seismic companies can be land, marine or active in both spheres. For example, Petroleum Geo-Services (PGS) is a purely marine seismic company, whereas IG Seismic Services provides only land seismic services. Large companies such as CGGVeritas and WesternGeco are active both on land and at sea.

It is difficult to assess the total number of seismic companies due to the high number of small firms in operation (Moore 2006). According to Veritas, there are minimal financial and technical entry barriers to the land seismic market. Veritas stressed the difficulty of discussing the global competition in land acquisition in any quantitative fashion due to the constantly changing configurations of seismic crews and the immense numbers of channels in the market (Veritas 2005). CGGVeritas confirmed the same market situation in 2011 by describing the nature of the land seismic market as fragmented and characterized by intense price competition in its low-end segment. According to the CGGVeritas 2011 annual report, the entrance into the international market of a significant number of formerly national competitors (such as Chinese BGP) drove prices down in 2011. Within the crowded environment of the land seismic market, the significant service providers are CGGVeritas, WesternGeco, Global Geophysical Services, BGP and Geokinetics (CGGVeritas 2011). They are involved in the low-end segment as well as the high-end segment which includes more complex land seismic acquisition projects, such as projects in desert and Arctic regions.

While the land seismic market is difficult to gauge, it is relatively easy to assess the number of companies involved in marine seismic operations because the number of vessels is known. Offshore's seismic vessel survey for 2011 counts 163 seismic vessels including the new builds[22] contracted for delivery but not yet in service (Kliewer 2011). The number of 3D seismic vessels, which are high-capacity vessels with six streamers[23] or more, was sixty-one at the end of 2010. The offshore sector has four leading participants: WesternGeco, PGS, Fugro and CGGVeritas. PGS is a marine-focused geophysical company and CGGVeritas is a pure-play geophysical company.

Aside from land and marine seismic companies, there are seismic equipment companies which specialize in manufacturing seismic signal sources (explosive or vibroseis sources), grids of sensors and recording systems. Traditionally, marine seismic equipment has held the majority share of the seismic equipment market at 55%, with surface equipment accounting for about 45%. Sercel (a division of CGGVeritas) and ION Geophysicals have the largest share of the global seismic equipment market. Their combined market share is about 80% (Deloitte 2010).

According to the World Oil Seismic Report published in 2006, WesternGeco was the largest seismic company followed by CGG, Veritas[24] and PGS, which are roughly all the same size in terms of equity market capitalization. The report indicates that these companies accounted for about 45% of the entire sector's earnings, which are estimated close to $5 billion. Outside these large contractors, small companies make up 80% of the remainder of the market and account for 55% of its total revenues. Revenues in Table 4.7 indicate figures for 2010: the revenue of four main companies accounts for

Table 4.7 Main Marine Seismic Companies

Main Marine Seismic Companies	Country of Origin	Fleet (2D & 3D)	Revenue (2014)
CGGVeritas	France	13	$3.09 billion
PGS	Norway	12	$1.45 billion
Fugro*	Netherlands	12	$1.56 billion
WesternGeco** (Schlumberger)	France / US	N/A	N/A

Source: Compiled by the author based on annual reports of each company and Offshore (2011b) & Offshore (2013)

*Note: * Fugro reported €2.57 billion annual revenue for all its activities. Oil and gas field activities contribute 80% of its annual revenue. For Table 4.7, 80% of 2.57 billion is taken and converted into USD at 0.76 €/$ (2014 annual average) exchange rate (Fugro 2014).*
***Schlumberger does not separately report WesternGeco's revenue or fleet size.*

$8.3 billion of an estimated $15 billion seismic market. The rest of the market belongs to smaller companies which become uncompetitive compared to the big four due to mobilization costs if survey locations are at distant sites. In addition, smaller companies are at disadvantage, being unable to offer a differentiating technology on which a premium price can be charged. The big three, WesternGeco, CGGVeritas and PGS, possess the Q-Marine single-point receiver system, BroadSeis and GeoStreamer respectively, promising improved imaging of complex geological environments such as pre-salt offshore Brazil, or comparable structures offshore West Africa where most of the lucrative seismic action is taking place. Their higher R&D spending, up to 5% of their turnover, has resulted in new technologies being owned by these large companies. Today, CGGVeritas is the biggest geophysical company by revenue, with $2.9 billion in 2010, followed by WesternGeco with $1.99 billion in the same year.

CGGVeritas: Compagnie Générale de Géophysique (CGG) was founded in 1931 in France while Veritas was established in 1974 in Canada. CGG and Veritas merged in 2005, creating the largest independent geophysical company. It is the world's leading international pure-play geophysical company delivering a wide range of seismic services and equipment via its equipment company Sercel to customers throughout the global oil and gas industry. Its geophysical services include offshore and onshore seismic acquisition, data processing and imaging and reservoir management. Sercel is the leading manufacturer of land and marine equipment, including high-tech integrated electronic recording systems, cables, land, ocean-bottom and borehole sensors, streamers, offshore seismic sources and vibrators. Processing and imaging services are conducted in open data-processing centres or its fifteen client-dedicated centres. Its reservoir software services company, Hampson-Russell, is used by over 500 oil and service companies worldwide. Following profits in 2007 (€250 million) and 2008 (€340 million), the

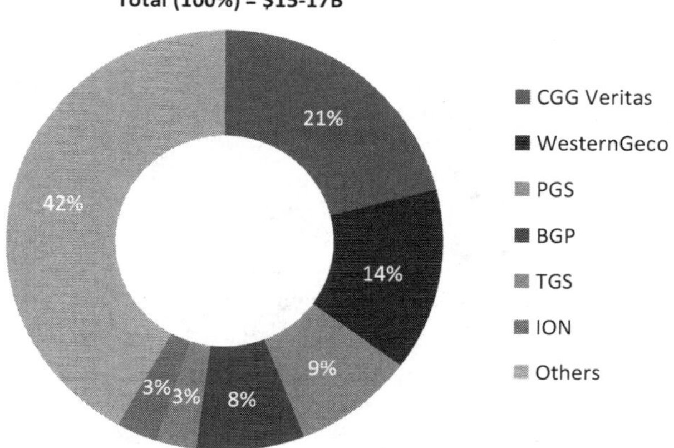

Figure 4.3 Main Seismic Companies

Source: Figure based on the information in Friedemann (2010, 7).

company has occurred losses during 2009 (€259 million), 2010 (€44 million) and 2011 (€9 million) (CGGVeritas 2011).

PGS: Petroleum Geo-Services ASA was founded in Norway in 1991. It became a marine-focused seismic company after selling its onshore seismic data acquisition business and onshore multi-client library business to US-based Geokinetics in 2009. PGS allocates seismic vessels between multi-client and single-client contracts (marine contracts). In a weak market, greater capacity is allocated to robust multi-client projects, whereas in a strong market PGS aims to maximize profits by executing more marine contracts. PGS is active in the high-density segment with its proprietary Geostreamer dual-sensor streamer technology. PGS spends more than $60 million on R&D annually. The company incurred a loss of $14 million in 2010 but made a profit of $33 million in 2011 (PGS 2011).

WesternGeco: WesternGeco is a Schlumberger company for geophysical services. It was formed in 2000 when Western Geophysical (a Schlumberger company) and Geco-Prakla (a Baker Hughes Incorporated company) merged to create a new joint venture incorporating the surface seismic data acquisition and processing businesses of the two companies. The new venture was known as Western-GECO and was owned 70% by Schlumberger and 30% by Baker Hughes (OFT 2000). In 2006, Schlumberger acquired a 30% minority interest in WesternGeco held by Baker Hughes, making WesternGeco a subsidiary of Schlumberger (Schlumberger 2012a). WesternGeco is a leading geophysical company offering a comprehensive range of seismic acquisition, data processing, reservoir imaging, monitoring and development services. WesternGeco acts in partnership with GeoSolutions (a software company) for seismic processing and for earth model building.

Table 4.8 Services Provided by Main Seismic Companies

	Data Library Sales	Data Processing	Land Equipment	Land Acquisition Services	Marine Equipment	Marine Acquisition Services
CCGVeritas	✓	✓	✓	✓	✓	✓
WesternGeco	✓	✓	Proprietary Q-Land	✓	Proprietary Q-Marine, Q-Seabed	✓
PGS	✓	✓	n/a	Divested Feb-2010 (Geo-Kinetica)	Proprietary	✓
BGP	n/a	✓	INOVA (51%)	✓		Announced Plans
TGS	✓	✓	n/a			✓
ION	✓	✓	INOVA (49%)		✓	

Source: Table based on the information in Friedemann (2010, 8).

WesternGeco Q Technology is a proprietary point-receiver acquisition system for enhanced reservoir description. Seismic interpretation is done by Schlumberger with Petrel seismic interpretation. WesternGeco generated $2.12 billion in 2009 with $326 million net income before taxes, and a revenue of $1.99 billion in 2010 with $267 million net income before taxes. Schlumberger integrated WesternGeco into its reservoir characterization team in 2011 and has since stopped publishing revenue and income data separately (Schlumberger 2010a).

Fugro: Fugro collects, processes and interprets data related to the earth's surface for clients in many sectors. Oil and gas field activities contribute 73% of its annual revenue (Fugro 2010). CGGVeritas agreed to buy Fugro's geosciences division for $1.6 billion in September 2012 to add four seismic survey vessels to its fleet and benefit from a forecast recovery in survey rates. The comprehensive deal includes the creation of seabed joint venture,[25] a commercial agreement for CGGVeritas to sell Fugro's existing multi-client data[26] and a global strategic, technical and commercial mutual preferred supplier agreement under which CGGVeritas and Fugro grant each other preferred supplier status for certain products and services required for the operation of their respective businesses. CGGVeritas praised the deal, stating that it allows access to high-end vessels at a time of market recovery. Fugro stated that it has decided to sell its seismic survey business because it lacks a market-leading position and earnings have been volatile. Fugro explained that 'this divestment allows Fugro to exit the capital intensive and volatile seismic segment of the oil and gas exploration market' (Bauerova 2012; CGGVeritas 2012a).

Geokinetics: Geokinetics acquired PGS's onshore seismic data acquisition and multi-client data library business in 2009, becoming a leader in the onshore seismic data acquisition business. The combination of Geokinetics and the onshore business of PGS has created the second-largest onshore seismic acquisition company in the world in terms of crew count, and the largest based in the Western Hemisphere (Oil Voice 2009). As a US-headquartered company, Geokinetics is active in unconventional shale plays in the US.

BGP: BGP is the biggest land seismic company headquartered in China and has long been responsible for the geophysical business of the Chinese state oil company (China National Petroleum Corporation, CNPC). BGP separated from CNPC as part of CNPC's long-term restructuring programme that has been spinning off its specialized service operations into separate companies (BGP 2012). BGP became a limited liability company after merging with six other Chinese geophysical companies in 2002. The company has stated that it held a 7.5% share of the world's geophysical market in 2003 (BGP 2003). ION Geophysical recorded that BGP held an 8% share in 2010 (Friedemann 2010). In 2010, ION Geophysical, based in North America and one of the world's largest seismic equipment manufacturers, and BGP, the world's largest land seismic company, launched a joint venture, INOVA Geophysical, that would focus on manufacturing land seismic equipment for the geophysical industry. BGP owns 51% and ION 49%, but INOVA has its own management team, its own locations and its own headquarters (ION 2010).

Relationship with International Oil Companies

Access to reserves is identified as the number one risk for oil companies, whereas frontier acreage[27] was ranked number one with regard to opportunities (Offshore 1996). Oil and gas companies are extending their search for new hydrocarbon resources into regions with deeper waters, harsher environments and more complex geologies, all factors which increase the need for seismic activity (Ernst & Young 2012). Advanced geophysical techniques such as 3D seismic, sub-salt imaging, electromagnetic and velocity modelling are used as means of generating accurate information from increasingly complex and remote reservoirs. Seismic data acquisition, processing and interpretation have gained tremendous importance in the world of international oil companies endeavouring to reach 'difficult-to-access reserves'. The two primary components of E&P decision making are seismic interpretation and reservoir modelling. The importance and sensitivity of seismic exploration for IOCs makes for an intense and profound relationship with geophysical services companies, on many fronts.

First of all, reservoir knowledge and the decision regarding where to drill are considered the key expertise of international oil companies. IOCs are determined to keep their essential reservoir expertise and consequently treat seismic exploration as a very sensitive segment. This reflects on the choice of which seismic exploration activities to be outsourced and which to be kept in-house. The majority of IOCs outsource the acquisition of seismic data entirely and its processing to a certain degree. They are, however, determined to keep the interpretation of the seismic data in-house.

Acquisition, in other words shooting surveys using highly specialized vessels, is left in the capable hands of seismic companies. On the processing side, although the policies vary, IOCs tend to outsource conventional and ordinary processing and do the complicated processing in-house. As for the interpretation, IOCs do it in-house due to the importance of seismic interpretation in making investment decisions. International oil companies identify seismic interpretation, geological interpretation and reservoir interpretation as three key in-house fields of expertise. IOCs have stated that they cannot take the risk of exploration and oil field development without having reservoir expertise in-house. By interpreting the data in-house, IOCs take on the liability of correct interpretation, deciding on the potential of oil-bearing formations before committing large sums of capital investment.[28] According to the results of the seismic interpretation, IOCs decide whether or not to drill exploration and development wells. Because the link between processing and interpretation is very close, IOCs tend to integrate interpretation during the treatment of data in order to avoid processing information that is insignificant or unrelated to the purposes of the company. If the processing is provided by a seismic company, IOCs may request that the person who will interpret the data be present during processing work. IOCs may also choose for the acquisition and processing of data to be done by different seismic companies.

In the case of joint ventures, JV partners decide whether one of the partner IOCs will treat the data or whether the processing will be outsourced to a geophysical services company. In these cases, IOCs may find themselves competing with seismic companies for the treatment of data in a project where partners decide which party will undertake the work. Some international oil companies wish to treat the data themselves in order to gain expertise in key areas such as deepwater; others do it as a default company policy. Interviewees stated that ExxonMobil processes seismic data in-house in parallel to third-party processing even if it does not win a tender. In fields considered very important, other IOCs may also choose to do the treatment themselves in parallel. For interpretation, a JV partner, generally an international oil company, is selected by JV partners. Other partners may do their own interpretation in parallel as a double-check.

Second, IOCs and OSCs work very closely on all seismic projects. During the acquisition, navigation of the ship, direction of travel and location of survey are predetermined by the customer. An IOC employee may be present on ship. Recorded tapes are sent to the seismic company's and IOC's offices. If the processing is done by the seismic company, the IOC will give precise specifications regarding the IT systems or algorithms to be used. Companies such as CGGVeritas have customer-dedicated processing centres located in their client's offices. These centres are responsive to the trend among oil companies to outsource processing work while maintaining a level of control. These dedicated centres ensure close coordination between oil companies and the seismic company and aid CGGVeritas to adapt its processing technologies to specific requests (CGGVeritas 2012b). During interpretation done in-house, IOCs use software developed by oil services companies. As in processing, very close cooperation in daily business takes place between IOC and OSC with an employee of the geophysical company resident in the IOC offices in order to consult on software applications. Anecdotal evidence suggests that Schlumberger's Petrel software is used by the majority of the oil and gas companies.

Third, IOCs guide technological progress in the seismic world in close cooperation with geophysical services companies, either by providing them with concepts for development or by inventing technologies which are passed on. Technology is exchanged continuously between international oil companies and seismic companies. For example, BP has led the development of 3D Wide Azimuth seismic methods in order to create sharper seismic images (BP 2012). Wide Azimuth technology has been developed by CGGVeritas at the request and prompting of BP. Another example that illustrates this two-way relationship can be found in the case of 3D seismic imaging technology, technology that changed the way the industry searches for oil and gas since it was invented by Exxon in 1963 (ExxonMobil 2007). 3D seismic surveys have since become a major industry-wide tool in exploration for hydrocarbons and have improved drilling success rates (Offshore 1995). A third example is provided by the Surface Slice Application,[29] which was developed by Exxon Production Research and incorporated into GeoQuest's IESX seismic interpretation software system (James et al. 1994). Externalization of technology from Exxon to Geoquest ensured that Geoquest would further develop the product and guaranteed reduced-fee access to Exxon. There are also many examples of the codevelopment of technology. The INTERSECT, a next-generation reservoir simulator, is one of the technologies codeveloped by Chevron and Schlumberger, combining Chevron's reservoir simulation capabilities and reservoir management experience with the leading software development capability and commercial experience of Schlumberger. Total also joined the collaboration at a later stage in order to contribute engineering resources and expertise to expand the INTERSECT simulator effort (Schlumberger 2012e).

Fourth, seismic companies work intensively with customers to develop new product lines or to provide services that respond to the most recent requirements of IOCs. OSCs' R&D groups aim to identify areas potentially interesting to customers, that reduce costs or that introduce a new attractive technology. Although NOCs are the primary customers of seismic companies, technological progress is driven by IOCs' requirements. According to the 2012 Seismic Vessel Survey, seismic companies are acquiring vessels to target the more difficult regions of the globe, including the Arctic and deeper water. Another trend is the increasing use of life-of-field seismic and ocean-bottom node installations for added geophysical insight into reservoir management (Kliewer 2011). These trends demonstrate that seismic companies respond to what international oil companies want to see in the future.

Finally, IOCs impact the industry structure by their actions, whether this is on a deliberate or an unplanned basis. An interviewee suggested that 'BP supported Veritas to create competition. Veritas was a technologically advanced company, but remained behind Western. BP made Veritas their preferred supplier and helped it grow. Wide Azimuth technology was developed by Veritas based on BP ideas. There was a clear desire from BP to help Veritas compete with Schlumberger'. Anecdotal evidence suggests that today other international oil companies are also seeking to favour the competitors of Schlumberger in order to form competition within the seismic business. An IOC suggested that it chose BGP instead of Schlumberger to avoid Schlumberger's dominating its supply of services.

4.3 WELL SERVICES

'Each oil well has its own personality.'

The broadest service segment within the industry is 'well services', also known as 'well support services' or 'oilfield services'; unlike other segments, which undertake a narrow, specific or highly specialized range of tasks, well services is wide ranging and may include diverse tasks varying from cementing to well testing. In the following paragraph, a more-detailed overview of the services which fall within this segment is provided.

An exploratory well is drilled following the results of seismic exploration. Drilling is the only sure way to confirm the presence of hydrocarbons, but simply drilling a hole into the ground rarely reveals conclusively whether it has intersected with a hydrocarbon-bearing zone. An exploratory well is just the beginning of a process that serves to acquire as much data and knowledge about subsurface structure and reservoirs as possible (Hermann et al. 2010). Wireline logging,[30] logging while drilling[31] and well testing[32] are all methods that further this understanding (Andersen 2011; Hermann et al. 2010). Furthermore, drilling—the exercise of creating a deep hole in the ground—requires complex services and technologies such as dynamic pressure management, drill bits[33] and drilling fluids[34] (Rigzone 2012a, 2012b). The complexities are evidenced in the report of the UK Parliament regarding the implications of the Gulf of Mexico oil spill. The report states that pressure in the well is controlled by ensuring that the pressure of the drilling fluid (mud) in the well bore, or bottom hole pressure, is sufficient to oppose the pressure from the oil, gas and water in the reservoir (the formation pressure). If the formation pressure is greater than the bottom hole pressure, oil and gas enter the wellbore, and may cause a blowout if not controlled. Tony Hayward, BP's former CEO, stated that 'the pressure on the drill pipe and the volume of drilling mud . . . are the two most important parameters that are monitored and measured on a continuous basis' (Tony Hayward quoted in UK Parliament 2011). The drilling fluid engineer monitors the formation pressure and adjusts the density of drilling mud to balance the pressure and keep the well bore stable (UK Parliament 2011). Controlling the pressure is not the only service needed during drilling. Throughout the construction of the well, services such as cementing, pressure pumping and stimulation are required. Additional technologies such as safety valves, packers and sand control technology are called for during the preparation of the well for production (i.e., completion). All these services, which carry out the majority of the tasks concerned with finding and extracting oil, are classified as 'well services'.

Well services are mainly categorized as falling within the 'drilling and evaluation' segment or the 'completion and production' segment. Drilling in the well services segment involves different services during the drilling process than those explained in section 4.2. In the drilling and evaluation segment, companies provide services for reservoir modelling, petro physical modelling, logging, drilling and precise wellbore placement solutions in order to model, measure and optimize well-construction activities. Companies also provide drill equipment components such as drill bits and the fluids necessary to lubricate and cool the drilling process. In the completion and production segment, cementing, stimulation, intervention, pressure control, speciality chemicals, artificial lift and completion services are provided. Companies provide equipment and chemicals necessary for completion, as well as the pump and monitoring systems for production.

The scope of this section is limited in order to provide a detailed explanation regarding all well services, which include a wide range of highly technical services requiring high-end technological solutions.

Four main companies offering well services have grouped their activities under certain categories. Analysis of the services indicates that well services often embrace the following activities:

Drilling & Evaluation Group:

- Drill bits; used for drilling, hole enlargement and coring.
- Drilling support services; directional and steerable drilling equipment and services.
- Well-construction services; casing running[35] and cementing[36] products and services (Rigzone 2012c).
- Wireline services; tools for open hole and case hole well logging used to gather data for petrophysical analysis, reservoir evaluation coring, logging while drilling and electric logging for reservoir and production flow evaluation, bottom hole sampling measurement while drilling systems.
- Well-testing services; specialized equipment and procedures to obtain reservoir information after the drilling has been completed. These include surface and downhole well-control equipment and services, including trees, surface test equipment, heaters, separators, burners and metering equipment and services. Well testing provides companies with vital information regarding the size and characteristics of the reservoir.

Completion & Production Group:

- Wellbore intervention services; products and services used in existing wellbores to improve their performance such as wellbore obstruction removal systems, thru-tubing fishing, electric and slick-line equipment and services.
- Artificial lift systems; systems such as hydraulic lift or gas lift that are established to lift hydrocarbons to surface for wells that do not have sufficient reservoir pressure.
- Completion services[37]; casing and cementing, tubulars, packers, safety valves, sand control technology, gas lift, electric submersible pumps, pressure control, speciality chemicals and accessories (Rigzone 2012c).
- Production enhancement services; such as advanced pressure pumping, chemicals, fracturing technologies, coiled tubing technologies provided to increase productivity and oil and gas recovery.

The demand for and supply of well services are linked to the level of expenditure of the oil and gas industry and requirements regarding exploration, development and production of oil and gas reserves. The growth of well services companies is a direct function of customers' exploration and production spending. Drops in oil prices impact revenues of well services companies whose business is based on complex equipment and services. Trends in oil exploration such as the rise of unconventional fields (tight oil and shale gas) also have an impact, because unconventional fields are highly capital intensive and require specific technologies such as horizontal drilling and hydraulic fracturing (fracking). While the cost of each service resists detailed explication, one interviewee mentioned that where a drilling rig in deepwater may cost around $0.5 million a day, drilling services may cost up to $1 million per day.

Market Structure and Main Actors

The well services sector is dominated by four main players: Schlumberger, Halliburton, Baker Hughes and Weatherford International. According to Morgan Stanley, after the purchase of Smith International by Schlumberger, the 'big four oligopoly' of oil services companies would control 75% of the market (Corkery 2010). On a market cap basis, Schlumberger, Halliburton and Baker Hughes are among the top five oil services companies. Schlumberger is the only one that is also involved in the geophysical services segment; the other companies focus entirely on well services. All companies offer well services such as drilling support services as well as oilfield products such as special chemicals, fluids and equipment. Each specific product line and service has its own leading company. However, interviewees mentioned that certain companies are better regarded than others in certain segments. Accordingly, Schlumberger is number one in drilling-related services whereas Halliburton is the leader in completion, stimulation and cementing and Baker Hughes is first in fluids.

The well services segment is characterized by the existence of small companies, four large market leaders and very few middle-sized companies. Small companies are established to develop new technology and acquired when they grow by large companies that wish to increase the span of their services. Schlumberger has bought nearly fifty small companies for their technology. One of the interviewees mentioned that 'Weatherford International escaped the net and has grown into a larger company'. A few examples of acquisitions are given below in the presentation of each company.

Schlumberger: Schlumberger is the world's largest oil services company with a market capitalization of $92 billion, which is higher than certain international and national oil companies such as Statoil, ENI and ConocoPhillips.[38] The Schlumberger brothers invented wireline logging as a technique for obtaining downhole data in oil wells, and Schlumberger was founded as a logging company in France in 1926 (Schlumberger 2012a). The company is active in two business segments: well services and seismic.

Table 4.9 Main Well Services Companies

Company	Headquarters	Revenue	Operating Income / Net Income	R&D Expenses	Leading Areas
Schlumberger	US–France	$48.6	$7.6 / $5.4	$1.2	Wireline logging, Well testing, E&P software, Directional drilling, Logging while drilling, Drilling fluids, Mud logging, Coil tubing, Stick line
Halliburton	US–Dubai	$32.9	$5.1 / $3.5	$0.6	Unconventional fields, Shale technologies, Deepwater technologies and production enhancement services in mature field, Cementing, Completion, Stimulation
Baker Hughes	US	$24.6	$2.9 / $1.7	$0.6	Tricone and PDC drill bits, Ream-while-drilling and casing drilling technology
Weatherford International	Switzerland	$14.9	$0.5 / $(0.6)	$0.3	Rotary steering systems, Artificial lift systems, Mature well extraction

Source: Compiled by the author based on companies' 2014 annual reports.

Note: All numerical data are in billion USD.

Schlumberger Oilfield Services provides oil companies with a wide range of well services, including but not limited to formation evaluation, directional drilling, well construction, completion, productivity and software and IT infrastructure. Oilfield Services manages its business through three groups: reservoir characterization, drilling and reservoir production. The segment accounts for over 90% of the company's revenue ($37 billion out of $39.5 billion in 2011). WesternGeco, Schlumberger's seismic business segment, provides advanced acquisition and data processing services.

Schlumberger is an industry leader in providing wireline logging, well testing, drilling and completion fluids, coiled tubing, measurement while drilling, logging while drilling, directional drilling services and mud logging, as well as fully computerized logging, geoscience software and computing services (Schlumberger 2011). It has a leading market share in the majority of its well services products. While no data is available, an interviewee stated that Schlumberger possesses a share of around 60 to 70% of the worldwide formation and evaluation segment, with its wireline logging market share at around 70%, and its logging while drilling technology market share standing at around 60%.

Schlumberger has launched itself as a technology provider and invests heavily in technology. The company spent $1.1 billion in R&D in 2011,

Table 4.10 Schlumberger Market Positions in 2010

	Service or Product	Spears Ranking
Reservoir Characterization Group	Geophysical Equipment & Services	2*
	Wireline Logging	1
	Production Testing	1
Drilling Group	Drill Bits	2
	Directional Drilling Services	1
	Rental & Fishing	3
	Drilling & Completion Fluids	1
	Logging while Drilling	1
	Surface Data Logging	1
	Solids Control & Waste Management	1
Reservoir Production Group	Pressure Pumping Services	2
	Completion Equipment & Services	4
	Artificial Lift	2**
	Coiled Tubing Services	1
	Specialty Chemicals	4

Source: Table based on the information in Gould (2011, 10).

*Note: *Schlumberger does not sell geophysical equipment. **Artificial lift ranking based on ESP market.*

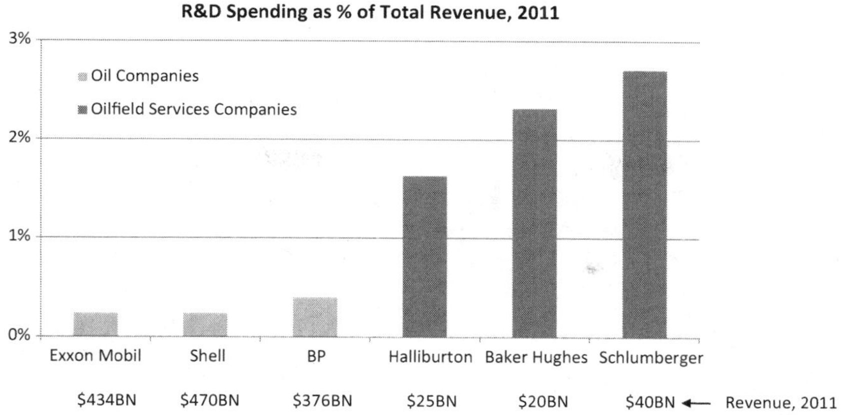

Figure 4.4 R&D Spending, 2011

Source: Based on the information in The Economist *(2012).*

which is more than all other well services companies combined and more than international oil companies' R&D spending as a percentage of total revenue (Economist 2012; Schlumberger 2012b).

During the interviews, Schlumberger's extensive range of services and its particular focus on technological knowledge were widely seen as its competitive advantage. The 'InTouch' knowledge system (an internal knowledge management system created by Schlumberger in 1998) has been cited as a competitive advantage by several industry experts. InTouch is a knowledge database via which users can connect operations knowledge on a worldwide scale. It is a key determinant of Schlumberger's current capitalization of knowledge and innovation. Thanks to the InTouch system, Schlumberger has been able to capture and share more knowledge and expertise company-wide, retain information when employees leave and develop a competitive advantage to improve its operations. The InTouch database contains more than one million knowledge items and is typically the first resort of field engineers experiencing a technical problem (Schlumberger 2012d). A Schlumberger employee can receive information from office headquarters and from experts around the world in answer to questions relating to operations via the InTouch system. For example, if an employee in Angola is having a problem with a logging tool, he is able to enter the problem in the InTouch system and leave an urgent message. The employee via his request is then immediately connected to a wide range of experts, greatly reducing the amount of time required to respond to and resolve operational problems. Schlumberger is recognized as the globally most-admired knowledge enterprise, thanks to its proprietary InTouch knowledge capture and sharing system (Schlumberger 2010b).

Another technology that is cited as an example of Schlumberger's competitive advantage is Petrel. This E&P software platform was acquired by Schlumberger in 2002 and has been developed and marketed as the only seismic-to-simulation software to provide all the tools an exploration company would choose to use. Petrel reduces the need for specific technologies for each segment of exploration, and allows users to interpret seismic data, build and simulate reservoir models and design development strategies in an integrated manner. Petrel allows companies to aggregate oil reservoir data from multiple sources, such as seismic acquisition and wireline loggings, and allows geophysicists, geologists and reservoir engineers to work in a collaborative manner via a single platform, and to integrate various operations through plug-ins. Thanks to Petrel, Schlumberger is able to offer a fully software-enabled workflow system, covering seismic and petrophysical interpretation as well as petroleum system modelling (Gould 2011). Throughout the years, Schlumberger has acquired several smaller size companies active in different aspects of exploration software and integrated them into the Petrel system. In 2009 it bought Techsia, a petrophysical software firm which had developed Techlog (a wellbore interpretation platform), and in 2010 it acquired IGEOSS (a developer of structural geology software); both products were subsequently integrated into Petrel (Schlumberger 2012a).

Schlumberger's history is full of acquisitions ranging from the above-mentioned small software-oriented acquisitions to large specific-technology company acquisitions. Schlumberger has been transformed from a logging company to a broad-range well services company mainly through acquisitions and R&D. The firm was engaged in several strategic acquisitions in 2010 and 2011. The $11 billion acquisition of Smith International, a company active in the design, manufacturing and provision of drill bits (a segment in which Schlumberger did not have significant operations), aimed to broaden yet again the services Schlumberger could offer. The biggest acquisition in the company history also gave Schlumberger complete ownership of MI SWACO, the drilling fluid business that Schlumberger and Smith had owned jointly since 1999. Its Smith acquisitions widened Schlumberger's lead over other well services companies (Casselman and McCracken 2010).

In 2010, Schlumberger also acquired Nexus Geosciences (a seismic support company), IGEOSS (a developer of structural geology software) and Geoservices (a well services company specializing in mud logging, slickline and production surveillance operations). Framo Engineering (a privately owned Norwegian firm producing pumps and metering systems) and Thrubit (a Shell Technology Ventures Fund company providing openhole logging services) were picked up the following year (Schlumberger 2012a). Strategic mergers and acquisitions remain central to Schlumberger's drive to offer a comprehensive package of products and services and build the most comprehensive service portfolio among OSCs (Schlumberger 2012a).

Halliburton: Halliburton is the second-largest well services company and third-largest oil services company. Founded in 1919 as a cementing

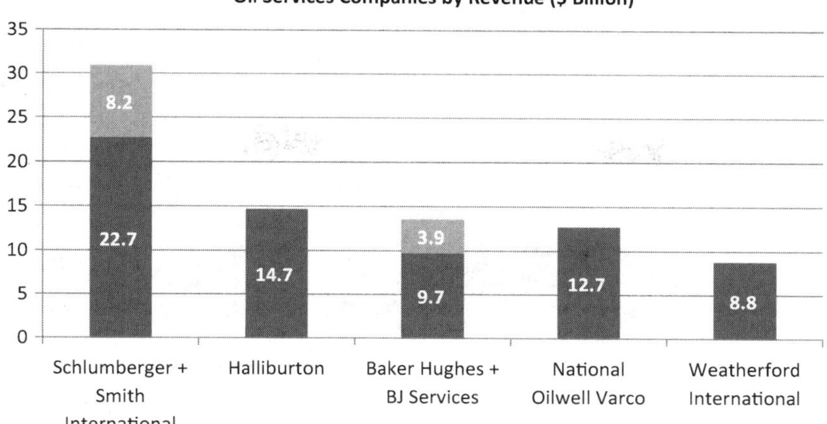

Figure 4.5 Oil Services Companies by Revenue

Source: Figure based on the information in Casselman and Mccracken (2010).

*Note: Including Mergers in 2009. *BJ Services fiscal year ended September 30. All figures are USD billion.*

company in the US, Halliburton has established a second headquarters in Dubai to strengthen its presence in the Eastern hemisphere. The company merged with Dresser Industries in 1998 in a $7.7 billion deal and became the largest OSC of that time, ahead of Schlumberger. Halliburton decided to focus on the well services segment and spun off its engineering and construction segment, KBR, in 2006. In 2011 Halliburton acquired Multi-Chem, a provider of production chemicals, to provide production assurance services in mature fields.

Halliburton focused its strategy on specific technologies required in unconventional, deepwater and mature fields. Halliburton is industry leader in unconventional shale plays and has developed in-house technologies that reduce completion times (Rapidsuite system) and improve environmental safety (Cleansuite system technologies) in hydraulic fracturing and water treatment during unconventional developments. In addition, it is constantly building up services and technologies such as high temperature and high pressure drilling for deepwater fields. When it comes to mature fields, Halliburton provides evaluation tools and technology allowing better characterization of the subsurface of these fields, often a first step in their redevelopment (Halliburton 2011).

Baker Hughes International (BHI): Baker Hughes is the third-biggest provider of well services, offering technologies and services to efficiently drill and complete wells. It was formed in 1987 by the merger of Baker International and Hughes Tool Company, both founded 100 years ago. Throughout its history, BHI has acquired and integrated numerous smaller size well services companies, such as Teleco (measurement while drilling and directional and horizontal drilling), Exlog (mud logging), Petrolite (speciality chemical), Centrilift (artificial lift) and Western Atlas (seismic exploration and well logging). Western Atlas was a large acquisition worth $6 billion and resulted in the creation of two new divisions, Baker Atlas (downhole, wireline, logging services) and Western Geophysical (seismic data services). In 2000, Western Geophysical joined with Schlumberger's geophysical operations division (Geco-Prakla) to form WesternGeco. Baker Hughes then sold its share to Schlumberger and quit the seismic business in 2006. BHI made another large acquisition in 2009, buying BJ Services for $5.5 billion. BJ Services was the third-biggest provider of pressure pumping services, used to stimulate production in conventional oil wells and to break up rocks in the shale formations of unconventional gas plays (Klamp and Lundgren 2009). Today, BHI organizes its business under three segments: a) drilling, evaluation and fluids; b) completions, production and chemicals; and c) pressure pumping, covering nine product line groups in total. BHI also offers reservoir development services (Baker Hughes 2011).

Weatherford International (WI): WI, the fourth-largest well services company, was formed by the merger of Energy Ventures and Weatherford Enterra in 1998. Like the three previously mentioned companies, WI has been involved in several acquisitions, including that of Precision Drilling Services, which was bought for $2.28 billion in 2005 in order for WI to increase its drilling services, wireline and evaluation capabilities. WI's areas of focus are drilling, evaluation, completion, production and intervention services. The company is particularly advanced in the completion and production segments. It is the only service company that provides all forms of artificial lifts and holds the record for the highest number of expandable sand screen systems (more than 600) installed worldwide (Weatherford International 2011).

Relationship with International Oil Companies

Well services is a technology-focused 'upstream' segment which is different from other segments, such as the manpower-concentrated drilling services or project management-oriented 'downstream' segments of EPC contractors. Services and products such as cementing or artificial lifts are highly specific and outsourced by nearly all international oil companies. In the past, IOCs undertook all well services themselves with the exception of wireline logging, which, according to an interviewee, was developed by Schlumberger and enjoyed protected status. Over time, IOCs have outsourced well services to advanced engineering companies that develop technologies that may overlap with the competences of international oil companies. Therefore the issue of competition between IOCs and OSCs usually arises with regard to well services firms—particularly Schlumberger and Halliburton—that offer integrated services. For example, Schlumberger

Oilfield Services offers its services through business models known as Integrated Project Management (IPM, for well construction projects) and Schlumberger Production Management (for field production projects). Under these models, Schlumberger offers complete projects rather than individual products and services. In certain IPM projects, Schlumberger charges a fee related to the oil, as in production enhancement projects, and may be paid in proportion to the amount of recovered oil. In its 2011 annual report, Schlumberger states that 'the projects may be fixed price in nature, contain penalties for non-performance and may also offer opportunities for bonus payments where performance exceeds agreed targets. In no circumstances does Schlumberger take any stake in the ownership of oil or gas reserves' (Schlumberger 2011). The fact that Schlumberger or Halliburton offer all-integrated projects to NOCs and charge a fee in the form of oil has caused competition worries to certain international oil companies which are specialized in integrated projects management and paid as a percentage of oil produced. As discussed previously, well services companies have stated during the interviews that they do not wish to compete with their customers and are instead focused on advancing technology.

International oil companies provide detailed specifications for the work executed by well service companies and supervise the work in progress. As it is the case in the seismic field, interpretation following logging or well testing is done by IOCs. An interviewee working in an oilfield company stated that 'international oil companies are in control of every decision. Everything goes through them; they manage the entire operation'. For example, in a logging service contract, the required logs, the programme of work and all other specifications are determined by the IOC. Well services companies provide the logging results to IOCs, which then interpret them to judge exploration risks. For each project, IOCs combine different technologies and services. They may require cementing and logging to be done or artificial lifts or drilling strings to be installed by different companies, aiming to mix the best or the most economical of all services.

International oil companies are experts in combining services and technologies. They participate in numerous joint research and development projects with various service companies in to order to improve the combined working of different services as well as their individual improvements. An example of a joint R&D venture is the Prometheus project run by Total and Halliburton. Logging while drilling (LWD) and rotary steerable technologies have traditionally been very difficult to deploy in hostile environments of high pressure and temperature because sophisticated tools are exposed to extreme heat downhole throughout the entire drilling operation. In 2008, Total entered into an agreement with Halliburton to jointly develop a suite of ultra-HPHT measurement and LWD sensors. In collaboration, the two companies are now developing the Prometheus suite of LWD tools specifically for the ultra-HPHT environment of some North Sea fields (Dirksen 2009).

R&D direction is steered by international oil companies' requests. Well services companies develop research and development programmes, form joint ventures or acquire specific capabilities in order to respond to IOCs' challenging requirements, such as drilling targets found at greater depths or with a more difficult well profile. The most recent example is the Schlumberger & Cameron joint venture in the subsea segment. Subsea production activities are especially challenging because they involve working with equipment on the ocean floor where it is subject to high pressure and temperature and strong ocean currents. Offshore services rates are higher due to these challenging conditions and the high level of service intensity, helping well services firms to increase their revenue per rig. Schlumberger and Cameron International Corp (provider of subsea equipment) signed an agreement in 2012 to jointly develop products, systems and services for the subsea oil and gas market under the joint venture called OneSubsea. Schlumberger CEO Paal Kibsgaard confirmed that the venture is intended to help customers improve production and recovery from their offshore wells. The subsea market is traditionally served by small and specialized firms, but because demand for subsea services is expected to increase, Schlumberger has been acquiring or forming joint ventures to strengthen its position in the market. It acquired Framo Engineering, a Norwegian company with technologies focused on the subsea market, in 2011, alongside forming a JV with Cameron (Team 2012). Subsea demand is created by both IOCs and NOCs such as Petrobras, but international oil companies are more active in the deepwater and ultra-deepwater fields. In addition, smaller companies that large companies later buy are often initially funded by international oil companies or start out as partners to IOCs. For example Schlumberger bought Thrubit, which had previously been funded by the Shell Technology Ventures Fund. IOCs give direction for technology development, while also helping small companies with seed money.

In addition to driving technology development, IOCs also impact on industry structure. When international oil companies find the market share of a particular company too high, they try to balance the situation by selecting other providers to work with. Some IOCs have complained that a limited number of companies exist which offer services with the required quality and health and safety controls for advanced technologies. The balance of power between international oil companies and well services companies has moved in favour of well services companies, especially when oil prices are high. It is IOCs' stated aim to generate competition in certain segments such as logging where Schlumberger nearly has a monopoly. A few interviewees stated that the relationship is like a 'nonformalized partnership'. There is permanent friction in the relationship, with up and down cycles.

Besides these impacts, there were a few issues mentioned as points to be improved by interviewees working for well services. First of all, several interviewees complained that IOCs outsourced specific technologies to the extent that they had lost the requisite knowledge to supervise the work. One of the interviewee stated, 'When you don't do the daily work, you forget the challenges that daily work presents'. Second, companies who have a particularly advanced technology in one segment stated that they do not appreciate being put in the same basket with other providers on the basis of a common denominator with minimum technology. They regret that they cannot always monetize the value and technological edge that they offer in a tender. Following the technical qualification based on minimum technical requirements, the choice is made by the procurement department based on price. Some well services companies do not feel that the 'cost of doing the work properly to the highest standards' is sufficiently valued. This is expected to change following the Macondo accident.

NOTES

1. The well services segment is mainly known as 'oilfield services' in the oil industry. This book uses 'well services' in order to avoid confusion between oil services companies (category name for all service companies involved in the oil field, ranging from seismic to engineering and construction services) and oil*field* services companies (companies working on the specific segment of well support services).
2. Logging refers to performing tests during or after the drilling process to allow geologists and drill operators to monitor the progress of the well drilling and to gain a clearer picture of subsurface formations.
3. A drilling rig is a machine used to bore radially into the ground, whereas an oil rig refers to the complex of equipment assembled to drill and service an oil well.
4. Jack-ups are bottom-supported units that do not float but stand on retractable legs. They can work in water depths that are shallower than the length of the legs. They are towed by towboats to their drilling location or are placed on heavy-lift vessels for transport over long distances. Floaters are mainly ships with drilling equipment. They can be designed as drillships or as semi-submersible units. Semi-submersibles do not rest on the sea floor like jack-up rigs but maintain their position over the well during drilling by using anchoring or dynamic positioning systems. They are suited to drilling in rough waters thanks to their stability. Drillships are vessels with drilling equipment installed on the deck. Drillships are especially useful for drilling exploratory wells thanks to their high mobility. Water depth ranges from 20 to 500 feet for jack-ups, whereas semi-submersibles or drill ships can be used in up to 12,000 feet of water (Diamond Offshore 2012; Hermann et al. 2010).
5. Few companies perform drilling and well servicing within the same structure. Generally, drilling companies (such as Transocean, Diamond Offshore) and well services companies (such as Schlumberger, Halliburton) are different companies. Well servicing companies are analysed in section 4.3.
6. In the oil industry, 'completion' means preparing the well for production. This entails running production casing, stimulation work and zonal isolation to allow the well to flow hydrocarbons (Diamond Offshore 2012).
7. The ranking by rig fleet shows differences in various compilations because some companies include platform rigs within their list while others do not.
8. The origin of the companies can be traced back to the Danciger Oil & Refining Company, which was founded in the US in 1919 and purchased its first

drilling rig in 1926. The current corporate structure belongs to the Offshore Company incorporated in the US in 1953. Southern Natural Gas Co. (SNG), a pipeline company, formed the Offshore Company to design and construct the world's first jack-up drilling rig and merged it with Danciger drilling operations in 1953. A year afterwards, Rig 51, the world's first offshore jack-up drilling rig, was developed by the Offshore Company in the US Gulf of Mexico. With oil exploration beginning to move further offshore, Offshore Company moved overseas and acquired International Drilling Co. Ltd. of London in 1963, marking the beginning of the company's operations in the North Sea. After going public in 1967, the company expanded its operations to Asia, where it drilled its first deepwater well. With the change of name from SNG to Sonat, the Offshore Company became known as Sonat Offshore Drilling Inc. (SODI). Sonat took advantage of investor optimism due to rising oil prices in 1993 and spun off SODI. In 1996, SODI acquired Transocean ASA of Norway for $1.5 billion, creating Transocean Offshore Inc., and doubling the company's offshore drilling fleet.

9. The Discoverer Enterprise is Transocean's first ultra-deepwater drillship with dual activity drilling technology; it can operate two drill operations at the same time and reduce the cost of an ultra-deepwater development project by up to 40%. It is almost as long as three US football fields (at 835 feet) and can drill a well more than 6.5 miles beneath its drill floor.
10. 10,000 feet is 3,048 metres and 1.894 miles; 12,000 feet is 3,660 metres and 2.27 miles.
11. Sedco Forex was formed by a merger of two drilling companies, the Southeast Drilling company (Sedco), founded in 1947 by Bill Clements and acquired by Schlumberger in 1984, and the French drilling company Forages et Exploitations Pétrolières (Forex) founded in 1942. Schlumberger purchased 50% of Forex in 1964 and merged it with 50% of Languedocienne to create the Neptune Drilling Company. The remaining 50% of Forex was acquired the following year; Neptune was renamed Forex Neptune Drilling Company.
12. The drilling company Reading & Beates tried to beat out Sonat Offshore in the Transocean ASA acquisition. Following its failure, Reading & Bates merged with Falcon Drilling Co. in 1997, then acquired Cliffs Drilling in 1998. Due to its high debt situation, its management decided the time was ripe to merge with Transocean Sedco Forex. Transocean acquired R&B Falcon for more than $9 billion in an all-stock transaction, which included the assumption of $3 billion in debt.
13. GlobalSantaFe is the product of the November 2001 merger of Global Marine Inc. and Santa Fe International Corporation. When assessed on June 18, 2007, the company's fleet consisted of forty-three cantilevered jack-up rigs, eleven semi-submersible rigs and three drillships, in addition to the two additional semi-submersible platforms it operated for third parties under a joint venture agreement.
14. In 2007, high commodities prices encouraged energy companies to ramp up their exploration efforts and created a shortage of rigs. That has led to a healthy backlog of business for companies such as Transocean, which supply rigs under multiyear contracts.
15. In 1988, Noble acquired six offshore rigs and twenty land rigs from Temple Marine Drilling and R.C. Chapman Drilling. Noble also acquired Peter Bawden Drilling Ltd. In 1991, Noble acquired five jack-ups and seven submersibles from Transworld Drilling. In 1993, Noble acquired nine jack-ups of the Western Oceanic fleet from The Western Company of North America and two submersibles from Portal Rig Corporation.
16. Noble sold land drilling assets and four posted barge rigs in 1996 and twelve mat-supported jack-ups in 1997.

17. In 2000, Noble formed a joint venture with Lime Rock Partners to acquire a harsh-environment North Sea jack-up named Noble Julia Robertson, and a further joint venture with Crasco to own and operate the Panon jack-up rig.
18. Its six drill ships and eight ultra-deepwater semi-submersibles average less than three years in age.
19. On land, the source of the 'bang' is either dynamite or a specialized truck called a vibroseis truck. The source is moved to different locations and all data from each geophone are recorded after each bang. At sea, marine vessels use a combination of airguns, waterguns and other acoustic sources to create the pulse needed to take seismic readings and record data by hydrophones. The procedure is essentially the same as on land except that the instruments are continuously moving. An array of airguns is towed behind the survey vessel and just below the sea surface. The airguns are fired at regular intervals as the vessel moves along predetermined survey lines. Energy reflected from beneath the seafloor is detected by numerous 'hydrophones' contained inside long, neutrally buoyant 'streamers' also towed behind the vessel.
20. There are two principal categories of seismic surveying. In two-dimensional (2D) seismic surveys, a single line of acquisition data is recorded, meaning that an interpretation is done on a single slice of the earth. In three-dimensional (3D) seismic surveys, multiple parallel lines of data are acquired, allowing a cube of interpreted data to be created. There is also 4D, which is mainly time-lapsed 3Ds, involving running the same seismic surveys several times—but this is not widely used. In summary, 2D seismic shows a single slice of the earth whereas 3D seismic shows a volume of the earth and 4D seismic shows a 3D volume at different times. 3D seismic is the main reference of data collection today for oil and gas exploration. Multi azimuth (MAZ) and Wide Azimuth are advanced acquisition techniques that involve shooting more than one energy source location (MAZ), or using a wide area of receivers by every shot point (WAZ). They are particularly useful in complex structures such as sub-salt areas.
21. TGS has no fleets of its own (no vessel on long-term commitment) and simply charters vessels short-term. It produced $170 million as net income in 2011.
22. Accordingly, the new vessels show a built-for-purpose trend in that many are being specified and outfitted for particular types of surveys such as 4D and ocean-bottom node applications.
23. A seismic streamer is a 'cable, trailed from a geophysical vessel, towing a series of hydrophones along the sea floor recording seismic signals from underwater detonations' (Glossary of Petroleum Industry 2015).
24. Before the CGG–Veritas merger.
25. Both companies agreed to create a joint venture (60% Fugro, 40% CGGVeritas) in seabed geophysics, which involves installation of permanent monitors on the ocean floor.
26. CGGVeritas will act as a nonexclusive broker of Fugro's existing multi-client library and receive commission fees on all multi-client sales. Data remains owned by Fugro.
27. Frontier zones are regarded as remote regions of the earth, or those with harsh climates or difficult-to-work-in environments such as deepwater or ultra-deepwater areas. They are areas in which little, if any, drilling and seismic acquisition have occurred (Offshore 1996).
28. Drilling an exploratory well can cost between $100 and $200 million.
29. Surface Slice interpretation allows interpreters to scan the 3D shape of the dome through horizontal slides that resemble a series of contour maps. It is a fast new volume interpretation tool.
30. Wireline logging involves lowering an instrument to the bottom of a well, then pulling it up slowly while recording the information provided by the

instrument. Founders of Schlumberger did the first well log in 1927 and measured electrical resistance of the earth by lowering an electrode on the end of a long cable and continuously recording the voltage difference while retracting it. Reservoirs bearing water or hydrocarbon reacted in different ways. The main wireline devices used today include several logging measurements such as gamma ray, resistivity, neutron and density. By comparing logs from many wells in the field, geologists and reservoir engineers can develop oil production plans (Hermann et al. 2010).

31. Logging while drilling tools acquire the same logging data (resistivity, sonic, nuclear) during the drilling itself (Hermann et al. 2010).
32. A well test involves setting up equipment to flow oil at a controlled rate through surface valves and then measuring flow rates, pressures, temperatures and the properties of the fluids produced to yield information on permeability, contents and potential flow rate of the reservoir (Hermann et al. 2010).
33. Drill bits are cutting tools that cut into the rock and create a cylindrical hole when drilling an oil or gas well. The drill bit is a rotating apparatus located at the tip of the drill string below the drill collar and drill pipe. Because different configurations work better on different formations, a number of drill bits may be used on one well (Rigzone 2012b).
34. Drilling fluids, also referred to as drilling mud, are added to the wellbore to facilitate the drilling process by suspending cuttings, controlling pressure, stabilizing exposed rock, providing buoyancy, cooling and lubricating (Rigzone 2012a).
35. 'The first step in completing a well is to case the hole. After a well has been drilled, should the drilling fluids be removed, the well would eventually close in upon itself. Casing ensures that this will not happen while also protecting the well stream from outside incumbents, like water or sand. Consisting of steel pipe that is joined together to make a continuous hollow tube, casing is run into the well' (Rigzone 2012c).
36. 'The next step in well completion involves cementing the well. This includes pumping cement slurry into the well to displace the existing drilling fluids and fill in the space between the casing and the actual sides of the drilled well' (Rigzone 2012c).
37. 'Well completion incorporates the steps taken to transform a drilled well into a producing one. These steps include casing, cementing, perforating, gravel packing and installing a production tree' (Rigzone 2012c).
38. Please see Tables 2.3 and 2.6 for further details on market capitalization.

REFERENCES

Andersen, M.A. 2011 (Spring). "Discovering the Secrets of the Earth, Defining Logging." *Oilfield Review Schlumberger* 23(1): 59–60.

API (American Petroleum Institute). 2011. "Explaining Exploration and Production Timelines (Offshore)." Accessed 8 April 2012. http://www.api.org/newsroom/upload/51073205_explaining_exploration_and_production_timelines_offshore1.pdf

Baker Hughes. 2011. "Well Planned, Managed, Delivered: 2011 Annual Report." Accessed 5 December 2012. http://phx.corporate-ir.net/phoenix.zhtml?c=79687&p=irol-reportsannual.

Baker Hughes. 2014. "Earnings from Innovation: 2014 Annual Report." Accessed 5 May 2015. http://phx.corporate-ir.net/phoenix.zhtml?c=79687&p=irol-reportsannual.

Bauerova, L. 2012. "CGGVeritas to Buy Fugro's Seismic Unit for $1.6 Billion." *Bloomberg* [Online], 24 September. http://www.bloomberg.com/news/2012–09–24/cggveritas-to-buy-fugro-s-seismic-unit-for-1–6-billion.html.

BGP. 2003. "CNPCs BGP Spinoff Expected to Capture Greater Land Seismic Market." *BGP* [Online], 7 April. Accessed 5 September 2012. http://www.bgp.com.cn/NewsIn.aspx?menu=11.

BGP. 2012. "About Us: Introduction." *BGP* [Online]. Accessed 5 September 2012. http://www.bgp.com.cn/09/About%20us/Introduction.htm.

BP. 2012. "Ground-breaking Technologies." *BP Subsurface* [Online]. Accessed 1 December 2012. http://www.bp.com/extendedsectiongenericarticle.do?categoryId=9037387&contentId=7068780.

Casselman, B., and J. Mccracken. 2010. "Schlumberger Deal Widens Oil-Services Lead." *The Wall Street Journal*, 22 February.

CGGVeritas. 2011. "Annual Report 2011, Form 20-F." Accessed 26 September 2012. http://www.cgg.com/InvestorArchivedReports.aspx?cid=4817.

CGGVeritas. 2012a. "CGGVeritas Announces Acquisition of Fugro's Geoscience Division and the Creation of Strategic Partnerships." CGGVeritas Press Release. 24 September. Accessed 30 October 2012. http://www.cggveritas.com/default.aspx?cid=5488.

CGGVeritas. 2012b. "Dedicated Processing Centers." Accessed 2 March 2012. http://www.cggveritas.com/default.aspx?cid=1944.

CGGVeritas. 2012c. "Seismic Overview." Accessed 2 June 2012. http://www.cggveritas.com/popup_page.aspx?cid=1–24–163.

CGGVeritas. 2014. "Annual Report 2014, Form 20-F." Accessed 26 May 2015. http://www.cgg.com/InvestorArchivedReports.aspx?cid=4817.

Clanton, B. 2010. "Pride Sees More Consolidation in Offshore Drilling Industry." *FuelFix* [Online], 4 November 2010. Accessed 5 December 2012. http://fuelfix.com/blog/2010/11/04/pride-sees-more-consolidation-in-offshore-drilling-industry/.

Corkery, M. 2010. "Big Oil (Services) Is About to Get Even Bigger." *The Wall Street Journal*, 19 February.

Daily, M. 2010. "Noble Corp to Buy Frontier Drilling for $2.16 Billion." *Reuters* [Online], 28 June. Accessed 5 September 2012. http://uk.reuters.com/article/2010/06/28/us-noblecorp-idUSTRE65R2C520100628.

Deloitte. 2010. "Seismic Equipment Market Review." Deloitte & Touche Regional Consulting Services Limited. Accessed 5 December 2012. https://www.deloitte.com/assets/Dcom-Russia/Local%20Assets/Documents/Energy%20and%20Resources/dttl_Seismic-Equipment-Market-Review_25072012_EN.pdf.

Diamond Offshore. 2012. "Offshore Drilling Basics." Accessed 2 September 2012.

Dirksen, R. 2009. "Hostile Drilling Environments Require New Approach." EPMAG, Hart Energy, 1 August. Accessed 15 December 2011. http://www.epmag.com/Production-Drilling/Hostile-drilling-environments-require-approach_42842.

Drilling Contractor. 2002. "Contractors Ordering Offshore Rigs on Speculation." *Drilling Contractor* May/June [Online]. Accessed 30 December 2012.

Economist. 2012. "The Unsung Masters of the Oil Industry." *The Economist*, 21 July.

Ensco. 2011a. "Annual Report and United Kingdom Statutory Accounts 2011." Accessed 5 September 2012. http://www.enscoplc.com/files/docs_financial/Ensco%20plc%202011%20Annual%20Report%20and%20UK%20Statutory%20Accounts.pdf.

Ensco. 2011b. "Ensco Plc to Acquire Pride International, Inc." Ensco Plc Press Release, 7 February. Accessed 16 August 2012. http://www.enscoplc.com/Newsroom/Press-Releases/Press-Release-Details/2011/Ensco-plc-to-Acquire-Pride-International-Inc1124029/default.aspx.

Ensco. 2012. "History: Ensco Rose from Humble Beginnings to Become the Leader in Customer Satisfaction and the Second Largest Offshore Drilling Company."

Accessed 1 December 2012. http://www.enscoplc.com/About-Us/History/default.aspx.

Ernst & Young. 2012. "Review of the UK Oil Field Industry." March.

ExxonMobil. 2007. "Analyst Meeting." 7 March. Accessed 1 June 2012. http://sec.edgar-online.com/exxon-mobil-corp/8-k-current-report-filing/2007/03/13/Section7.aspx.

Friedemann, C. 2010. "Seismic Sector Overview." *ION Geophysical, Investor Education Series* [Online], May 2010 Accessed 22 May 2015. http://www.iongeo.com/investoreducationcenter/PDFS/IR_SeismicSectorOverview_Friedemann_PP_100802.pdf.

Fugro, N.V. 2014. "Annual Report 2014." Accessed 15 June 2015. http://www.fugro.com/docs/default-source/investor-publications/2014/annual-report-2014.pdf?sfvrsn=10.

Funding Universe. 2012. "Transocean Sedco Forex Inc. History." *Funding Universe* [Online]. Accessed 1 June 2012. http://www.fundinguniverse.com/company-histories/transocean-sedco-forex-inc-history/.

Glossary of Petroleum Industry. 2015. "Seismic Sea Streamer in English", *Babylon*, http://translation.babylon.com/english/seismic+sea+streamer/.

Gould, A. 2011. "Schlumberger CEO Presentation." 27th Annual Sanford C. Bernstein Strategic Decisions Conference, Schlumberger Limited. http://www.slb.com/news/presentations/2011/~/media/Files/news/presentations/2011/20110602_agould_sanford_bernstein.ashx.

Halliburton. 2011. "Advancing Technology, Delivering Results: 2011 Annual Report." Accessed 5 December 2012. http://ir.halliburton.com/phoenix.zhtml?c=67605&p=irol-reportsAnnual.

Halliburton. 2014. "Go Big: 2014 Annual Report." Accessed 5 May 2015. http://ir.halliburton.com/phoenix.zhtml?c=67605&p=irol-reportsAnnual.

Hermann, L., E. Dunphy, and J. Copus. 2010. *Oil & Gas for Beginners: A Guide to the Oil Industry*. Global Markets Research. London: Deutsche Bank.

Hewitt, T. 2008. "Who Is to Blame ? Allocating Liability in Upstream Project Contracts." *Journal of Energy & Natural Resources Law* 26(2): 177–184.

Hill, A. 2011. "BP's Woes Are a Guide to Modern Executives." *The Financial Times*, 18 January. http://www.api.org/newsroom/upload/51073205_explaining_exploration_and_production_timelines_offshore1.pdf.

Hyperdynamics. 2010. "Hyperdynamics Awards 3D Seismic Processing Contract." *Hydrodynamics Corporation Press Release* [Online], 13 July. Accessed 16 August 2012. http://www.slb.com/news/press_releases/2012/2012_0726_intersect_collaboration_pr.aspx.

IHS. 2015. "IHS Petrodata Weekly Rig Count." Accessed 26 May 2015. https://www.ihs.com/products/offshore-oil-rig-data.html.

ION. 2010. "ION/BGP JV Is Off to a Fast Start." *ION Geophysical* [Online], October. Accessed 15 December 2012. http://www.iongeo.com/content/documents/pdfs/articles/EP_INOVA_Fast_Start_101014.pdf.

James, H., M. Tellez, G. Schaetzlein, and S. Stark. 1994 (July). "Geophysical Interpretation: From Bits and Bytes to the Big Picture." *Oilfield Review Schlumberger*, 23–31.

Kaiser, M.J., and B.F. Snyder. 2012. "Reviewing Rig Construction Cost Factors." *Offshore Mag*, 1 July: 72.

Klamp, E., and K. Lundgren. 2009. "Baker Hughes to Buy BJ Services for $5.5 Billion." *Bloomberg* [Online], 31 August. Accessed 8 August 2012. http://www.bloomberg.com/apps/news?pid=newsarchive&sid=a0dd4ngIH_t8.

Kliewer, G. 2011. "Seismic Vessel Count Remains Steady." *Offshore Mag*, 1 March: 71.

Krauss, C. 2010. "A Behind-the-Scenes Firm in the Spotlight." *The New York Times*, 25 May.

Malek, C., L. Hermann, and J. Copus. 2009. "Chasing the Pendulum." Global Markets Research. London: Deutsche Bank.

McBarnet, A. 2011. "Leap of Faith for Dolphin." *Oil Online* [Online], 6 September. Accessed 5 September 2012. http://www.oilonline.com/default.asp?id=259&nid=19274&name=Leap+of+faith+for+Dolphin.

McBarnet, A. 2012. "Call for Better Margins." *Petroleum Review* [Online]. Accessed 5 September 2012. http://content.yudu.com/Library/A1vx02/PetroleumReviewMarch/resources/16.htm.

McFarland, J. 2009. "How Do Seismic Surveys Work?" *Oil & Gas Lawyer* [Online], 15 April. Accessed 8 April 2012. http://www.oilandgaslawyerblog.com/2009/04/how-do-seismic-surveys-work.html.

McKinsey.1997. "Alliances in Upstream Oil and Gas", *The McKinsey Quarterly*, Number 2, David ERNST, Andrew M.J. STEINHUBL

Moore, J. 2006. "Seismic Market Report: Issues and Trends." *World Oil*, November.

National Commission. 2011. "Report to the President: Deepwater, the Gulf Oil Disaster and the Future of Offshore Drilling." National Commission on the BP Deepwater Horizon Oil Spill and Offshore Drilling, January.

Natural Gas. 2011. Exploration. *NaturalGas.Org* [Online]. http://www.naturalgas.org/naturalgas/exploration.asp.

NBC News. 2007. "Oil Drillers Transocean, GlobalSantaFe to Merge." *Associated Press* [Online], 23 July. Accessed 5 September 2012. http://www.msnbc.msn.com/id/19911184/ns/business-oil_and_energy/t/oil-drillers-transocean-global santafe-merge/.

Noble. 2011. "Towering Above: 2011 Annual Report." Accessed 5 September 2012. http://www.noblecorp.com/assets/Docs/AR11/NE-AR2011.pdf.

Noble. 2012. "Towering Above: A Brief history of a Great Company." Accessed 1 December 2012. http://www.noblecorp.com/assets/flipbooks/Noble-History/index.html.

NY Times. 2007. "Transocean and GlobalSantaFe to Merge." *The New York Times* [Online], 23 July. Accessed 5 September 2012. http://www.bloomberg.com/news/2011–08–15/transocean-reports-all-cash-voluntary-offer-to-buy-aker-drilling.html.

Offshore. 1995. "Exploration 3D Seismic Boosting Wildcat Success, Reducing Well Count." *Offshore Mag*, 1 April: 55.

Offshore. 1996. "Frontier Exploration: Remote and Deepwater Frontier Areas Seeing More Exploration." *Offshore Mag*, 1 April: 56.

Offshore. 1999. "Driller Consolidation Begins but Will It Continue." *Offshore Mag*, 8 January: 59.

Offshore. 2000. "Drilling Rig Contractor Mergers May Not Be Over Yet." *Offshore Mag*, 1 October: 60.

Offshore. 2007. "Top 10 Drilling Contractors: Rig Utilization Stands at 100% Around the Globe." *Offshore Mag*, 1 February: 67.

Offshore. 2008. "US Gulf Rig Market Wanes; Rest of World Prospers." *Offshore Mag*, 1 February: 68.

Offshore. 2009. "Rig Market Adjusts to Economy, Oil Price." *Offshore Mag*, 1 February: 69.

Offshore. 2010. "Top 10 Offshore Drilling Contractors: Modest Recovery Possible in Offshore Rig Market." *Offshore Mag*, 1 February: 70.

Offshore. 2011a. "Rig Market Review: Reviewing the World Offshore Rig Market." *Offshore Mag*, 1 February: 71.

Offshore. 2011b. "Worldwide Seismic Vessel Survey." *Offshore Mag*, March [Online]. Accessed 19 December 2012. http://www.offshore-mag.com/content/dam/etc/medialib/platform-7/offshore/maps-and_posters/1103off-vessel-survey.pdf.

Offshore. 2012. "Top 10 Drilling Contractors: Drilling Contractors Ready Fleet for Upcoming Activity." *Offshore Mag*, 1 February: 72.

Offshore. 2013. "Worldwide Seismic Vessel Survey." *Offshore Mag*, March [Online]. Accessed 19 June 2015. http://www.offshore-mag.com/content/dam/offshore/print-articles/Volume%2073/03/0313-seismic-vessel-survey.pdf

OFT (Office of Fair Trading). 2000. "Completed Merger of the Surface Seismic Data Acquisition and Data Processing Interests of Schlumberger Limited and Baker Hughes Incorporated." Accessed 20 March 2011. http://www.oft.gov.uk/OFTwork/mergers/mergers_fta/mergers_fta_advice/schlumberger.

OGJ. 1999. "Sedco Forex Offshore to Merge with Transocean." *Oil & Gas Journal* 97(29).

OGP & IAGC (International Association of Oil & Gas Producers & International Association of Geophysical Contractors). 2011. "An Overview of Marine Seismic Operations."

Oil Voice. 2009. "PGS to Sell Onshore Seismic Business to Geokinetics." *Oil Voice* [Online], 3 December. Accessed 15 December 2010. http://www.oilvoice.com/n/PGS_to_Sell_Onshore_Seismic_Business_To_Geokinetics/f8a503a2d.aspx#ixzz2CyA8QOMn.

PGS. 2011. "Annual Report 2011." Accessed 15 July 2012. http://www.pgs.com/pageFolders/40374/Annual_Report_2011.pdf.

Reinhardsen, J.E. 2012. "President & CEO Presentation." Pareto Securities Oil & Offshore Conference, 13 September.

Reuters. 2011. "Hercules Offshore, Inc. Signs Purchase Agreement to Acquire Assets of Seahawk Drilling." *Reuters* [Online], 11 February. Accessed 5 September 2012. http://www.reuters.com/article/2011/02/11/idUS228941+11-Feb-2011+PRN20110211.

Reuters. 2012. "Shell Admits Arctic Drilling Defeat, for Now." *Reuters* [Online], 17 September. Accessed 20 October 2012. http://www.reuters.com/article/2012/09/17/royaldutchshell-idUSL1E8KHHA120120917.

Rigzone. 2012a. "How Do Drilling Fluids Work?" *Rigzone* [Online]. Accessed 30 September 2012. http://www.rigzone.com/training/insight.asp?i_id=291.

Rigzone. 2012b. "How Does a Drill Bit Work?" *Rigzone* [Online]. Accessed 30 September 2012. http://www.rigzone.com/training/insight.asp?i_id=326.

Rigzone. 2012c. "How Does Well Completion Work?" *Rigzone* [Online]. Accessed 30 September 2012. http://www.rigzone.com/training/insight.asp?i_id=326.

Rigzone. 2012d. "Offshore Rig Day Rates." *Rigzone* [Online]. Accessed 30 December 2012. http://www.rigzone.com/data/dayrates/.

Rigzone. 2012e. "Rig Report: Offshore Rig Fleet by Manager." *Rigzone* [Online]. Accessed 30 September 2012. http://www.rigzone.com/data/rig_report.asp?rpt=mgr.

Schlumberger. 2010a. "Full Year 2010 Results." Accessed 26 September 2012. http://investorcenter.slb.com/phoenix.zhtml?c=97513&p=irol-resultsNewsArticle&ID=1518462&highlight=.

Schlumberger. 2010b. "Schlumberger Cited for Knowledge Management." *Schlumberger News* [Online], 3 December. Accessed 16 August 2012. http://www.slb.com/news/inside_news/2010/2010_0312_make_award.aspx.

Schlumberger. 2011. "Annual Report 2011." Accessed 26 September 2012. http://investorcenter.slb.com/phoenix.zhtml?c=97513&p=irol-reportsannual.

Schlumberger. 2012a. "About Schlumberger: History." Accessed 16 May 2012. http://www.slb.com/about/history.aspx.

Schlumberger. 2012b. "Company Information: Backgrounder." Accessed 16 May 2012. http://www.slb.com/about/who/backgrounder.aspx.

Schlumberger. 2012c. "Diagram of Rig Types and Operating Environments." *Oilfield Glossary* [Online]. Accessed 16 May 2012. http://www.glossary.oilfield.slb.com/en/Terms/r/rig.aspx.

Schlumberger. 2012d. "Knowledge Management." Accessed 16 August 2012. http://www.slb.com/services/westerngeco/services/dp/people/km.aspx.

Schlumberger. 2012e. "Total Joins Chevron and Schlumberger Collaboration on Development of the INTERSECT Next-Generation Reservoir Simulator." *Schlumberger Press Release* [Online], 26 July. Accessed 16 August 2012. http://www.slb.com/news/press_releases/2012/2012_0726_intersect_collaboration_pr.aspx.

Schlumberger. 2014. "Annual Report 2014." Accessed 26 May 2015. http://investorcenter.slb.com/phoenix.zhtml?c=97513&p=irol-reportsannual.

Sea Drill. 2015. "Fleet Status Report." *Sea Drill* [Online], 6 May. Accessed 5 June 2015. http://www.seadrill.com/~/media/Files/S/Seadrill/our-fleet/sdrl-fleet-status-q4–2014.pdf.

Strahan, A. 2007. "Transocean to Acquire GlobalSantaFe for $17 Billion (Update7)." *Bloomberg* [Online], 23 July. Accessed 23 July 2007. http://www.bloomberg.com/apps/news?pid=newsarchive&refer=home&sid=avjX0oYyowNA.

Team, T. 2012. "What the Cameron Subsea Deal Could Mean for Schlumberger." *Forbes* [Online], 27 November. Accessed 2 December 2012. http://www.forbes.com/sites/greatspeculations/2012/11/27/what-the-cameron-subsea-deal-could-mean-for-schlumberger/.

TGS. 2011. "Annual Report 2011, See the Energy." Accessed 26 September 2012. http://www.tgs.com/uploadedFiles/Investor_Relations_Zone/Form_-_Annual_Report/AR2011FINAL.pdf.

Total. 2012. "Geophysics Context." *Total Technohub, Total's Exploration & Production Techniques Magazine*.

Transocean. 2007. "Transocean Inc. and GlobalSantaFe Corporation Agree to Combine." *Transocean Press Release* [Online], 23 July. Accessed 23 July 2007. http://www.slb.com/news/press_releases/2012/2012_0726_intersect_collaboration_pr.aspx.

Transocean. 2011. "Proxy Statement and 2011 Annual Report." Accessed 5 September 2012. http://media.corporate-ir.net/media_files/irol/11/113031/AR-2011/HTML2/default.htm.

Transocean. 2012. "Our Company: Our History." Accessed 1 December 2012. http://www.deepwater.com/fw/main/Our-History-3.html.

Udall, R., and S. Andrew. 2008. "The Offshore? Good Luck, Bad Luck and Mukluk." *Energy Bulletin* [Online], 11 September. http://www.energybulletin.net/stories/2008–09–11/offshore-good-luck-bad-luck-and-mukluk.

UK Parliament. 2011. "Challenges of Deepwater Drilling." 6 January. Accessed 20 March 2011. http://www.publications.parliament.uk/pa/cm201011/cmselect/cmenergy/450/45005.htm#n1.

Veritas. 2005. "Annual Report 2005, Form 10-K." Accessed 26 September 2012. http://www.cggveritas.com/data/1/rec_docs/109_Annual_Report_2005.pdf.

Visiongain. 2011. "The Advanced Oil & Gas Exploration Technologies Market 2011–2021." 22 September. Accessed 25 November 2011. https://www.visiongain.com/Report/687/The-Advanced-Oil-Gas-Exploration-Technologies-Market-2011–2021.

Weatherford International. 2011. "Annual Report 2011." Accessed 5 December 2012. http://www.weatherford.com/weatherford/groups/web/documents/weatherfordcorp/annualreport2011.pdf.

West, J., A. Walker, and M. Pickup. 2011. *Offshore Rigs: Is the Next Wave of Consolidation on the Horizon?* Equity Research. London: Barclays Capital.

Wethe, D. 2012. "Transocean Beats Analyst Expectations with Rig Cost Control." *Bloomberg* [Online], 2 August.

Wethe, D., and M. Stigset. 2011. "Transocean Offers Double Market Value for Aker Drilling." *Bloomberg* [Online], 15 August. Accessed 5 September 2012. http://www.bloomberg.com/news/2011–08–15/transocean-reports-all-cash-voluntary-offer-to-buy-aker-drilling.html.

5 Conclusion

As illustrated throughout the book, the oil industry functions on the basis of a nexus of contracts between national oil companies, joint venture partners, international oil companies and oil services companies and their subcontractors. Because the focus of the book has been on international oil companies and oil services companies, the concluding section will analyse the deductions made regarding these two groups of companies.

The main question the book addresses is to what extent and why the relationship between IOCs and OSCs has changed in recent years and whether this change has impacted the nature of the firm in the oil industry. The findings reveal three categories of results: robust conclusions, weaker conclusions and inconclusive results.

The first robust conclusion concerns vertical disintegration and the resulting increase in the coordination function of the IOCs. Advancements in technology and IT capability alongside decreasing financial returns witnessed in the '90s as a consequence of low oil prices have increased outsourcing in the oil industry. The earlier model where IOCs executed projects in-house has evolved towards a vertically disintegrated model in which oil services companies provide technologies and services under the management of IOCs. As a result, international oil companies have increased their conscious coordination across the value chain and currently fulfil the functions of a 'core system integrator'.

This function performed by the core system integrator should not be confused with the integration of upstream and downstream value chain segments. While it is true that international oil companies have often been called 'integrated oil companies' because they are present all along the oil supply chain—from exploration to marketing of refined products—the integration function here consists of integrating the supply of services and technologies provided by oil services companies in the upstream segment of the oil value chain. In other words, not only are IOCs integrated companies present in upstream and downstream segments, but they also act as core system integrators combining various technologies and services within the upstream segment.

Following the vertical disintegration, IOCs increased their capacity to understand the pros and cons of each service, technology and supplier in order to integrate them successfully. They combine all required technology for exploration, drilling, well testing, well completion, construction of production platforms and subsea systems and manage all activities as one big project. Taking the characteristics of each oil project into account, IOCs decide on the combination of services required and ensure the coordinated functioning of all oil service companies involved in the field. In other words, IOCs work as the core system integrators, coordinating and integrating diverse activities in the oil field provided by numerous OSCs. They have become experts in finding the best combination and optimum structure for oil field development.

A second robust conclusion concerns the transformation in the structure of the oil services industry. IOCs' subcontractors have increased their capabilities, either by staying small, specialized companies or by growing and extending their offer to a wider range of services. The oil services industry structure has become polarized and is marked by the coexistence of large and small companies and the decreasing share of midsized companies. The move towards outsourcing, cost-cutting pressure from IOCs and the discovery of oil in increasingly complex fields requiring large, highly advanced contractors have caused several subcontracting companies to merge and consolidate. The 'cascade effect' has been observed across the value chain. In most service areas large OSCs have emerged, such as CGGVeritas and WesternGeco in seismic, Transocean and Noble in drilling and Schlumberger and Halliburton in well services. The large service companies have been growing annually by around 10% on average. In 2012, oilfield services companies grossed around $750 billion according to McKinsey consultancy (Economist 2012). Simultaneously, small companies focusing on niche technologies continue to exist, until they grow to midsize companies and are acquired by large OSCs. Whereas there is still a fair amount of competition in most parts of the industry, with each large OSC having different strengths and numerous smaller ones occupying specialized niches, the demand for complex services outstrips the supply in technically challenging and geographically remote areas such as ultra-deepwater (Economist 2012).

A third robust conclusion concerns the evolution in the relationship between the IOCs and the OSCs. In an industry where a field development can take over a decade from discovery to production, and where technical challenges increase day by day, profound and intricately connected relationship between contractors is inevitable. Although tenders with arm's length prices appear to be the norm, the relationship between international oil companies and the oil services companies extends far beyond the arm's length price relationship.

To begin with, OSCs shape their services in relation to the requirements of IOCs. As such, Transocean has been realigning its fleet to cater to the growing deepwater market where mainly IOCs are active. In 2012, the company announced the sale of thirty-eight shallow water rigs while simultaneously signing a ten-year contract with Shell to provide it with four newly built ultra-deepwater drilling ships (Team 2012).

Furthermore, OSCs develop technologies and direct their research and development programmes in close consultation with the requests or projected needs of international oil companies. R&D direction is heavily influenced by the IOCs, either by the ideas that they develop or by their requirements for advanced technology to find and extract oil under more challenging conditions. OSCs invest in technologies based on trends determined by international oil companies. Technology is exchanged continuously between IOCs and OSCs. For example, BP initiated the development of 3D Wide Azimuth seismic methods, which were subsequently developed by CGGVeritas at the request of BP (BP 2012). Moreover, the three-dimensional seismic imaging technology that changed the way the industry searches for oil and gas was first invented by Exxon in 1963 and further developed by seismic companies (ExxonMobil 2007). Another example is service companies that develop equipment for harsher conditions such as HPHT. In 2008 Total entered into an agreement with Halliburton to jointly develop a suite of ultra-HPHT measurement and LWD sensors. In addition to providing overall direction, IOCs support R&D in OSCs by participating in joint research projects or by providing seed funding to small technology companies. From being self-centred and internal, R&D in OSCs has been evolving towards an outward-looking model where IOCs participate in collaborative research and financing. The financial participation and research collaboration of international oil companies in R&D is important for OSCs as it shows IOCs' commitment to their relationship and ensures the development of products required by the end-customer.

In addition to influencing R&D, international oil companies impact the industry structure of oil services companies. This can either be intentional, for example where IOCs deliberately support the competitor of a strong OSC in order to prevent the OSC from becoming a monopoly, or unintentional. An instance of the latter case can be observed in the consolidation caused by IOCs through the price pressure they exercised in the '90s; another unintended impact they had on the sector was caused by their requirement for substantial and advanced services. IOCs' need for sophisticated companies that can offer services with the requisite quality and health and safety standards for sizeable projects in technically challenging areas has led to the development of large OSCs such as Schlumberger, Halliburton and Transocean. OSCs have been involved in several mergers and acquisitions in order to develop the required capabilities. For example Schlumberger has acquired over thirty companies in the past ten years alone in order to grow its portfolio of services. Another example can be seen in the Acergy and Subsea 7 merger that took place in 2010 and was completed in order to compete with larger size companies such as Technip and Saipem. During the merger, both companies declared that 'the new firm will be better able to meet the growing size and technical complexity of subsea projects, driven by the demand to access ever more remote reserves in increasingly harsh environments' (Goldstein 2010). Hence IOCs have an impact on the industry structure of oil services caused both by their deliberate actions and by the trends they initiate.

Moreover, international oil companies intensively manage the day-to-day operations of OSCs during the execution of services. Because each oilfield development is unique, IOCs determine all the specifications in detail and control appropriate execution of day-to-day activities. Employees of both IOCs and OSCs are present on the oil fields and often work within the same premises. International oil companies set the standards of business and require all services partners to adhere to them. For example, to underline its commitment to safe operations and environmental prudence, Shell has told OSCs, 'If you choose to break the rules, you choose not to work for Shell' (Wetselaar 2012).

A final robust finding concerns the increased coordination function exercised by the IOCs. Their role as core system integrator, changes in the service sector structure as well as the transformation of the relationship between IOCs and OSCs have all impacted the nature of the firm in the oil industry. The traditional clear boundaries that existed in the past between IOCs, OSCs and even NOCs have been blurred. Firm boundaries have been challenged with a strong drive towards collaboration, increased coordination and deep integration in day-to-day activities as well as in long-term business strategy. Instead of managing internal production factors, the core firm—in the oil industry the IOC—is now 'managing' production factors provided by external firms, namely oil services companies. OSCs develop R&D programmes, form joint ventures or acquire specific capabilities in order to respond to international oil companies' challenging requirements, such as meeting drilling targets in deeper areas with more complex well profiles. OSCs also work under the strict direction of IOCs on the field, with job specifications and scope defined by IOCs. As we have seen in the Macondo accident analysed in chapter 3, although the legal proceedings have not yet concluded, the responsibility and liabilities on an oil field remain with the operator, in our case the IOC.

There are two weaker conclusions. The first concerns the assessment of the degree to which the oil industry is following the same pattern as other industrial sectors vis-à-vis the management of the supply chain by the core system integrator. The second concerns the cyclical nature of the relationship between IOCs and OSCs.

The answer to the question of whether the oil industry is following the same patterns as other industrial sectors is yes, as discussed previously. The numerous similarities include consolidation among the services sector and the IOCs, codevelopment and cofinancing of R&D, the provision of overall technical direction by the IOCs, the closeness with which IOCs and OSCs work together, the centralization of procurement and the monitoring of each element of the supply chain from cost minimization to technical requirements.

On the other hand, however, there are specific limits within the oil industry regarding the coordination function of the international oil companies. First of all, international oil companies are locked into certain rules on how to tender and negotiate by the host country, NOCs or joint venture partners. In most cases, rules are clearly prescribed. IOCs' selection of suppliers is limited by tender rules or other requirements of JV partners and NOCs, such as local content requirements or the necessity of having a minimum number of participants in tenders.

Second, in the oil industry supply chain the end product is oil, which is not a manufactured product similar to an automobile or aeroplane. The cost of parts is essential in the manufacturing industries because they are the main components of the assembled product. The cost of oil services companies in exploration and production projects is equally vital, as it is the primary expenditure of international oil companies. However, the expenditure itself is not the main driver that determines the profitability of an oil field. The value of each field is derived from the amount of oil found and its global market price, both of which are uncertain. In addition, the system integrator deals with much higher uncertainty in the oil industry compared to other industries. In manufacturing industries, the system integrator is responsible for planning and assembling production. The risk that the system integrator takes is small, and uncertainty relating to the end product diminishes with each production step. In the oil industry, the system integrator plans and develops the field while also bearing the uncertainty of the subsurface. There is always a high degree of uncertainty about the profitability of any given oil field. This level of uncertainty regarding the end product does not exist in many other industry sectors. The role of the system integrator in the oil industry involves not only integrating the value chain, but also taking the risk of the unknown. Moreover, the majority of equipment and services in the oil industry are customized and dedicated to a specific project. Because each field development is unique, mass production is very limited. The technology and services required by an IOC need to be customized for each specific oil field development. Each technology and service embodies new concepts and technical progress. R&D in the supply chain is vital because the progress of technology and equipment is critical to the success of finding oil and increasing the amount of oil that can be recovered. The customized production of technology and services is different from the case of the automobile and aerospace industries, where the main company repeatedly produces the same product. In the oil industry, while there are pieces of equipment used in each field development such as pipelines, valves, drillships, etc., most services and equipment are customized.

Finally, in other large industries such as the automobile and aerospace, there are one or two main integrating companies. For example, in the aerospace industry, most suppliers work for Boeing or Airbus. In the oil industry, oil services can be provided to numerous companies including international oil companies but also independents or national oil companies. Due to these differences, there are inherent limitations to the coordination function of the core system integrator.

When it comes to cyclical relationship, the majority of industry experts agree that the relationship between IOCs and OSCs along the supply chain has been very cyclical. The balance of power between IOCs and OSCs shifts depending on oil price. Oil prices and fluctuations in the world economy have had the strongest influence over the cyclical relationship. High oil prices intensify the search for oil and the development of new fields, thereby increasing the workload of OSCs and reducing the available capacity. When oil prices are high, OSCs can demand higher prices for their services and negotiate better contractual terms with IOCs. The sharing of risks and profits moves in favour of OSCs. Conversely, when oil prices are low, IOCs have less interest in evaluating and exploring new oil fields, which reduces the demand for OSCs' services. Resulting overcapacity of services and equipment enables international oil companies to negotiate better terms with OSCs and move the risk–profit balance in their favour. Nearly all interviewees stated that oil price movement is the main factor impacting the relationship between IOCs and OSCs; despite this, very limited evidence is publicly available. There is no observable decrease of joint R&D projects during periods of low oil prices, and specific contract clauses related to risk sharing between IOCs to OSCs remain confidential.

Finally, the inconclusive findings can be grouped under two headings: the decreasing importance of IOCs and the degree of outsourcing. The first deals with the increasing importance of NOCs at the expense of IOCs. Robin West, chairman of PFC Energy, an industry consultancy, says: 'The reason the original Seven Sisters were so important was that they were the rule makers; they controlled the industry and the markets. Now, these new Seven Sisters [referring to NOCs] are the rule makers and the international oil companies are the rule takers' (Hoyos 2007). Their decreasing access to reserves and limited control over oil production has put IOCs at a disadvantage compared to NOCs in terms of reserve holdings, pushing IOCs towards developing giant and technically challenging oil and gas fields and affecting their relationship with OSCs (Jaffe and Soligo 2007). However, whether this has increased or decreased their impact on OSCs is debatable. While OSCs now work on a higher number of projects with NOCs, IOCs continue to be their key customers. Because IOCs typically manage larger and more technically challenging projects, they require greater use of OSC services. Therefore, IOCs continue to significantly influence technological progress in the oil services industry.

Second, there is continuous soul searching regarding the types of activity being outsourced and the degree of vertical disintegration within the industry. IOCs are in consensus regarding keeping certain activities of the supply chain in-house and ensuring vertical integration. Some interviewees argue that IOCs are at the limit of transferring responsibility and have become too dependent in OSCs. Accordingly, the oil business is growing dependent on advanced technologies that can only be provided by OSCs. Outsourcing has left IOCs hostage to the availability of increasingly expensive and sought-after services, which can only be provided by certain number of OSCs (Economist 2012). Other interviewees, on the other hand, consider that the business and risks taken by each sort of company are very different and do not see any danger as long as IOCs retain the ability to combine the best technologies and continue to take the risk of the reservoir. The findings are, therefore, inconclusive regarding this matter.

In a nutshell, the relationship between IOCs and OSCs provides an example of deep integration by core system integrators and offers insights into the change in the nature of the firm in the era of modern globalization. The challenges ahead make the traditional model of procuring each service without long-term supply chain management even more obsolete. Present trends suggest that in the future there will be even deeper supply chain integration in the oil industry. All companies need to develop their business models accordingly.

REFERENCES

BP. 2012. "Ground-Breaking Technologies." *BP Subsurface* [Online]. Accessed 1 December 2012. http://www.bp.com/extendedsectiongenericarticle.do?categoryId=9037387&contentId=7068780.

Economist. 2012. "The Unsung Masters of the Oil Industry." *The Economist*, 21 July.

ExxonMobil. 2007. "Analyst Meeting." 7 March. Accessed 1 June 2012. http://sec.edgar-online.com/exxon-mobil-corp/8-k-current-report-filing/2007/03/13/Section7.aspx.

Goldstein, S. 2010. "Acergy, Subsea 7 Rally on $5.4 Billion Merger." *The Wall Street Journal*, 21 June.

Hoyos, C. 2007. "The New Seven Sisters: Oil and Gas Giants Dwarf Western Rivals." *The Financial Times*, 12 March.

Jaffe, A.M., and R. Soligo. 2007. "The International Oil Companies." The James A. Baker III Institute for Public Policy, Rice University. Accessed 24 January 2009. http://www.bakerinstitute.org/programs/energy-forum/publications/energy-studies/docs/NOCs/Papers/NOC_IOCs_Jaffe-Soligo.pdf.

Team, T. 2012. "Shell's Record Transocean Deal Shows the Importance of Ultra-Deepwater." *Forbes* [Online], 10 February. Accessed 2 December 2012. http://www.forbes.com/sites/greatspeculations/2012/10/02/shells-record-transocean-deal-shows-the-importance-of-ultra-deepwater/.

Wetselaar, M. 2012. "Global Opportunities, Local Approaches." *Trends and What They Mean for Contractors*. Deloitte—Dutch Oil & Gas Conference 2012.

Appendix: Research Methodology

The structure of the oil industry and the relationship between international oil companies and oil services companies are constantly evolving, with many variables changing over time. It is impossible to handle intertwined variables that interact with each other via experiment, survey or archival analysis alone. In order to achieve the intended result, exploring the industry and business phenomenon requires a complete research strategy involving observation, interviews, archival analysis and company-based research. Case study is selected as the most suitable method because 'a case study is an empirical enquiry that investigates a contemporary phenomenon within its real life context, especially when the boundaries between phenomenon and context are not clearly evident' (Yin 2003, 13). Furthermore, a case study is appropriate when 'a "how" and "when" question is being asked about a contemporary set of events, over which the investigator has little or no control' (Yin 2003, 9). Case study results are given as examples in chapter 4.

The book questions how, why and to what extent the oil industry has changed. Focus is placed on the relationship between international oil companies and oil services companies, excluding national oil companies and independents. Because the oil services industry is very wide, three segments have been selected to analyse the relationship: offshore drilling services, seismic exploration services and well services.

A wide scope of literature ranging from books and articles to relevant news, papers and company-specific data has been analysed. News and special commission reports have been the best source for recent events such as BP's Macondo accident, whereas company annual reports and industry-related papers have been particularly useful in the analysis of segments in case studies.

In order to develop a deep understanding of the subject, semi-structured interviews were conducted not only with employees within the main international oil companies and oil services companies but also with people working in national oil companies or independents and individuals conducting industry-related research within universities, special research companies and investment banks. All thirty-nine interviewees were chosen on the grounds of their deep understanding of and insight into the subject and their key positions within the industry. The interviewees hold a diverse set of positions within sixteen different organizations ranging from top management, such as CEOs and CFOs, to engineers and managers who work directly on oil fields. This has ensured the collection of broad and overall strategy-oriented views, as well as in-depth, practical and hands-on observations and opinions. Thirty-three of the interviews were conducted face-to-face in London, Paris, Houston and Cambridge, with six being conducted over the phone due to the tight schedules of participants. Six different conferences and speeches were also attended during the research period.

REFERENCE

Yin, R.K. 2003. *Case Study Research: Design and Methods*. London: Sage.

Index